The *Earliest Record* of *Beekeeping* in *Northern England*

by Robert J Hawker

Northern Bee Books

The Earliest Record of Beekeeping in Northern England
© Robert J Hawker

All rights reserved. No part of this publication may be reproduced, stored in a retrieval system, transmitted in any form or by any means electronic, mechanical, including photocopying, recording or otherwise without prior consent of the copyright holders.

ISBN 978-1-908904-70-6

Published by Northern Bee Books, 2015
Scout Bottom Farm
Mytholmroyd
Hebden Bridge HX7 5JS (UK)

Design and artwork by D&P Design and Print
Printed by Lightning Source UK

The *Earliest Record* of *Beekeeping* in *Northern England*

by Robert J Hawker

Contents

Preface		3
Chapter 1	Introduction	5
Chapter 2	Historical context	7
Chapter 3	Northumberland Beekeeping and the Civil War	13
Chapter 4	A Treatise of Bees	23
Chapter 5	Development	31
Chapter 6	The Place of 'A Treatise of Bees'	47
Chapter 7	Time Line	53
Chapter 8	Concluding Observations and Unresolved Issues	55
Appendix I	Bonham catalogue entry	59
Appendix II	Some relevant extracts from the University of Sheffield's digitisation of Samuel Hartlib's papers	61
Appendix III	Firr deals	65
Appendix IV	The octagonal bee-hive	67
Appendix V	Miscellaneous	87
References		89

The Earliest Record of Beekeeping in Northern England

Newcastle upon Tyne in 1600. By permission of Newcastle upon Tyne Libraries.

Preface

All collectors of books on bees and beekeeping will be very aware of entry number 30 in the International Bee Research Association's *British Bee Books, A Bibliography 1500 – 1976,* **ARM, F**. *A Treatise of Bees,* 1646 -58. Whilst none of us are ever going to own this manuscript, I consider myself very privileged to have had the opportunity to study such a document written by an individual under considerable stress at the time and in very difficult conditions, about the most relaxing and rewarding of occupations - beekeeping. My sincere thanks go to The Beinecke Rare Book and Manuscript Library of Yale University, America, its Curator of Early Modern and Osborn Collections, Kathryn James, and her staff for their efforts in providing an electronic copy of the manuscript for me to study: also to Kathryn's predecessor, Stephen Parks who 'lit the fuse' to this a mere twelve years ago.

This dissertation has had some of the creases ironed out of it during talks delivered to groups of beekeepers, and has benefitted by assistance from some of the individuals who listened to them. Particular mention is deserved for my wife, whose help is always forthcoming and who continues to be very tolerant and supportive of my unusual hobby. Jeremy Burbidge proprietor of Northern Bee Books, also deserves a special mention for his encouragement and for not having the profit motive as his only purpose in life.

I make no apology for re-visiting some well documented areas of beekeeping history where I felt it necessary to do so, in order to fix the position of the MS in relation to known published material of the period. Also there is always the possibility that new information might yet emerge. I do not pretend that I think that this is the finished project: several times during my research new facts came to light which modified some of my earlier conclusions, and I accept that there might still be some 'loose ends', but 'time and tide etc.'.

I also do not apologise for using the Imperial system for measurements throughout: it was the values as given in the literature. For those so inclined the www provides a large number of sites to convert to metric values. To give the metric equivalent in the text every time a measurement is quoted, I believe is a distraction.

To any 'proper' historian who encounters this work, I apologise for the much abbreviated history in Chapter 1, but I write as a beekeeper primarily for beekeepers.

This is a story of a journey that includes joy, disappointment, experimentation, discovery, destruction, devastation, and satisfaction, played out to a backdrop of religious differences and intolerances, political upheaval, plague, pestilence, civil war and regicide; but mostly it is about beekeeping. For me a major challenge was to discover the author, and in the usual telling of all mysteries, I do not reveal his name until the end, necessitating the very annoying use of X until then. However, unlike all the best proponents of the art of mystery story-telling, I was unable to contrive an assembly of all the suspects in one place to deliver the reveal. Not surprisingly his name is not ARM. F.

Chapter 1

Introduction

In the early part of the 20th century, a manuscript that had resided in the library of a large Northumberland residence and belonged to the descendants of its author, is believed to have come into the possession of a very prominent member of the British Beekeepers Association at the time, it then passed through the hands of at least two British bee book collectors. By the late 1960s it was appearing in the lists of Bernard Quaritch, a major and well respected London dealer in rare books and manuscripts, but by 1971 it had transferred to the lists of fledgling book dealers, Hofman and Freeman. It was bought from their list by Torge S. K. Johansson, biologist, bee expert, scholar, writer and gentleman. Joseph Bray, generally acknowledged as Americas leading dealer in antiquarian bee books, very highly rated bibliophile and historian was the next to own it. His initial offer to sell it to Harvard University was declined, and Joe then offered it to Yale University where it has remained to this time in the James Marshall and Marie-Louise Osborn collection of the Beinecke Rare Book and Manuscript Library.

Some twelve years ago I was approached through intermediaries, by the then curator of the Beinecke library, enquiring if I knew who the author of the 17th century MS was. He identifies himself in the manuscript as F. ARM, but this I believe gives only an oblique clue as to his identity. Many hours studying local records failed to provide the answer, and I returned to researching and writing *The Enigma That Was Thomas William Cowan*. With that finished I returned to the quest to discover the identity of the 17th century Northern bee author. However, I soon realised that this was a distraction; it was more important to determine whether the manuscript had anything to offer by way of beekeeping techniques of the period and whether it modified the existing perceived beekeeping history, before worrying about who wrote it.

The Beinecke Rare Book and Manuscript Library very kindly provided an electronic copy for me to study. The condition of the manuscript appears to be very good for its age, but it was written using a quill, on both sides of very thin paper under challenging conditions; the ink was probably carbon black and one can only speculate as to the liquid carrier! For most of the treatise the author uses normal 17th century English but he is not averse to interjecting some local 17th century dialect; punctuation appears to have been optional for him, and he frequently uses different spellings for the same word, even on the same line. He writes in a reasonably readable hand, but there are occasions when the ink has not transferred to the paper or he has just omitted the odd letter or more. All adding to the fascination and frustration.

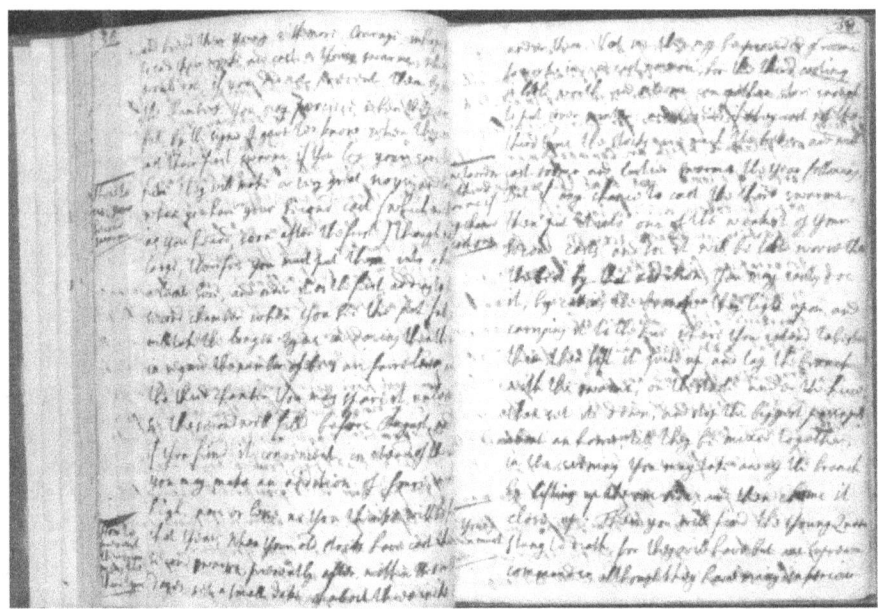

Copy of page from manuscript.

Today there are several new copies of beekeeping manuals published each year as every beekeeper feels he or she has something new to contribute to the age old craft, thus justifying yet another 'how to keep bees' publication. Often the only novel part is the title, which of course, has to be different. In the first half of the 17th century most authors of bee books, concerned themselves with the habits of bees rather than step by step, season by season, instructions on how to deal with them. This manuscript was unusual but not unique for the period because it is a beekeeping manual, dealing with the new system for keeping honey-bees in his unique and somewhat unorthodox hives. The instructions would have been welcomed and necessary for those who wished to follow his system of keeping bees in his hives. His offering was not just a regurgitation of old material: present day authors please note!

The order of material leaves a lot to be desired, and the repetition was not helpful, with each phrase taking time to decipher and interpret. I have declined to simply reproduce the treatise – even if Yale University would have consented to it, and the resulting short dissertation is my interpretation of X's treatise. I leave it to the reader to decide if it was time well spent, I enjoyed the journey: of the identity of X, more later.

Chapter 2

Historical Context

The Stuart period in England ran from 1603 until 1714; it was one of the most disturbed periods in the history of Great Britain. It had something of most things; assassination attempts, civil war, regicide, usurpation, national disasters, revolution etc. all under the rule of the Scottish Royal House of Stuart.

The previous monarch, Elizabeth I, had been in position since 1533, never married and hence did not produce an heir, making her the last monarch of the Tudor dynasty. In 1570 the Pope declared Elizabeth illegitimate, and excommunicated her, releasing her subjects from their allegiance to her. This led to the Catholics being banned from openly practising their religion in England; they were forced to attend Anglican services and punished if they did not. Elizabeth died in 1603 and was succeeded by James VI of Scotland who became James I of England, uniting the crowns of the two Countries; he was the son of Mary Queen of Scots whom Elizabeth had tried for treason and subsequently had executed in 1587. English Catholics thought that James, who was married to a Catholic, would be more sympathetic and allow them to openly practise their religion. That belief was reinforced when Thomas Percy at the behest of the Earl of Northumberland, a Catholic sympathiser, went to Scotland on a fact finding mission and reported back that James was likely to be so inclined. Initially James I lived up to the Catholics hopes, repealing the recusancy fines and promoting some Catholic sympathisers to high office. As would be anticipated this displeased the 'other side', especially when the Catholics became stronger and started to pose a threat. James, unable to satisfy the demands of the Puritans, turned against the Catholics and reinstated the situation as under Elizabeth I. He had experienced the same problems that had previously beset his mother in Scotland.

Not all Catholics were prepared to accept this and some were determined to do something about it, and as always in such situations a natural leader appears, in this case Robert Catesby. He met with kindred spirits Thomas Wintour, Jack Wright and Thomas Percy in May 1604 to decide on a course of action. The group leased a house in Westminster, recruited explosives expert Guy Fawkes, who had honed his skills in the Spanish Army, and installed him as caretaker in the house under the alias of John Johnson, illustrating the depth of thought being exercised. The 'gunpowder plot' was underway; Parliament and King were to be blown up!

However, the meeting of Parliament, an essential element in the plan, was continually delayed. The tightly bound group used the delay to recruit further members; Robert Keyes, John Grant, Kit Wright, Robert Wintour, all related in some way, and Catesby's servant Thomas Bates, all joined the original five. The group of conspirators had always known that simply blowing up the King and Parliament was unlikely to achieve the desired result, and some form of uprising immediately following was also necessary. For this the leaders required money and assets, so they also recruited Ambrose Rookwood, Francis Tresham and Sir Edward Digby. This was to prove their downfall.

The group had leased a cellar directly under the House of Lords into which Fawkes had moved a large amount of gunpowder and means of igniting it, all ready for November 5th when Parliament was to be eventually opened. Catesby, Wright and Bates had already left for the Midlands where they were to lead the uprising there and Digby was in place to kidnap the daughter of King James, Princess Elizabeth, with a view to installing her as Queen in exile on the Continent. However, Francis Tresham had written to his father-in-law Lord Monteagle, telling him to stay away from the opening of Parliament; Monteagle took the letter to Lord Salisbury who ordered the Parliamentary buildings to be searched. Fawkes and the gunpowder were found; he was quickly linked to a conspiracy plot - helped by him being spotted on one of his many trips to the Continent. He was arrested and tortured. The remainder fled on horseback and met up with Catesby, Wright and Bates, the combined group eventually making their way to Holbecke House in Staffordshire which they felt could be defended, but Grant had been blinded drying his gunpowder and the High Sherriff of Worcester easily took the castle on the 8th of November, killing Catesby, the Wright brothers and Percy in the process. Thomas Wintour, Rockwood and Grant were captured and it was not long before Digby, Robert Wintour, Keyes, Tresham and Bates joined them in the Tower of London. Tresham died in the Tower, Digby pleaded guilty, Robert and Thomas Wintour, Bates, Grant, Rockwood, Keyes and Fawkes were tried and found guilty of treason. They were all hung, drawn and quartered on 30th and 31st of January.

Prior to his trial Bates admitted that he had given details of the plot to Father Tesimond a Jesuit priest, thus dragging them into the conspiracy. Tesimond had passed the information to Henry Garnet the senior Jesuit through the sanctity of the confession, excusing him from reporting it. Garnet was executed whilst Tesimond escaped to the continent. The Earl of Northumberland was imprisoned in the Tower until 1621. Monteagle was awarded an annual pension for his passing of the letter.

The gunpowder plot was a failure and the perpetrators had been swiftly dealt with but the consequences for the Catholics would take centuries to overcome, all generally forgotten when the event is celebrated with bonfires and fireworks every 5th November. It clearly gave strength to James I for the rest of his reign and an excuse for boys of all ages to celebrate once a year.

James died in 1625. He was succeeded by his second son Charles who was 25 at the time. Charles became heir at age 12 following the death of his brother. He was inculcated in the profound belief that Kings were appointed by God and hence governed by divine rule; his reign reflected this. He was reserved, diffident and lacked self-assurance all poor characteristics for a monarch. At the very beginning of his reign English Protestants were calling for intervention on the Continent against Spain and the Catholic powers in the religious wars raging in Europe. Lord Buckingham was in charge of Charles's foreign policy and with flawed logic, embarked upon expeditions against Spain and France, both disastrous. Twice Parliament tried to impeach Buckingham resulting in Charles dissolving Parliament on both occasions. In 1628 Buckingham was 'assassinated'. Without Parliament Charles had no access to funds for more overseas diversions. When re-convened Parliament soon passed the *Petition of Rights* which, amongst many things meant Parliament had to approve any taxes being levied. Parliament next turned their attention to the religious policies of Charles: Charles responded in 1629 by dissolving Parliament for the third time. This began what was to become eleven years of rule without Parliament; eleven years of tyranny.

It began well, the economy prospered and Charles made peace with France and Spain. However, without Parliament his only access to funds was via some rather dubious new 'taxes' he introduced. But the religious problems had never been resolved and were unlikely to go away. In 1633 Charles appointed William Laud as Archbishop of Canterbury, who set out to restore the wealth of the Church and with Charles's support impose uniformity of worship throughout the Kingdom. As would be expected not all factions would be pleased, the strongest opposition coming from the Puritans, (effectively zealous Protestants), who felt that Laud's doctrines were too close to Catholicism. Laud used all means to silence his critics and pursue his ambition to introduce a common liturgy throughout the three kingdoms. The Scots reacted violently to the Book of Common Prayer and Arminium liturgy being forced upon them; the Scottish National Covenant was the outcome, formally accepted in Scotland in 1638. The Covenanter movement became the dominant political as well as religious force in Scotland and was in direct opposition to the King, making some sort of clash inevitable, Charles personality meaning that compromise was never a possibility. The Covenanter Army initially concentrated on defeating the last of the Royalist strongholds in Scotland. Charles went to York in March 1639 in an attempt to raise a Northern army, but progress was slow and not aided by the opposition of the English Puritans. Eventually the two opposing armies assembled on their respective sides of the border: negotiations to avoid all-out war took place in Berwick upon Tweed, Northumberland, in June 1639. Both sides were in fear of the others strengths, and a settlement was reached which neither side were happy with and unlikely to implement. This is generally known as the first Bishops War.

Charles was aware that his army was under strength, but he still believed that he could beat the Scots into submission. Yet again he re-assembled Parliament (the short Parliament), to raise taxes and went around Europe with the begging bowl, neither successful, but he re-assembled what there was of his army in Yorkshire and Northumberland. They were ill trained, lacking in discipline, poorly armed, unpaid and underfed, and had not been joined by an Irish army as Charles had planned. In August 1640 the Covenanter army again massed on the border. Viscount Conway, third choice as leader of the Royalist Northern Army, put most of his efforts into defending Berwick. On the 20th August General Leslie crossed the Tweed[1] at Coldstream, by-passed Berwick and headed straight for Newcastle upon Tyne. The Scots army arrived at Newburn, West of the City on the 27th, having marched along the Tyne thus avoiding the heavily defended Northern side of Newcastle. The Scots made a cavalry advance the next afternoon that was repelled, but the artillery duel that followed was a complete mis-match and the Royalist army retreated, Conway accepting that they were in no shape to defend Newcastle. They arrived in Durham very early the next day. Leslie marched unopposed into Newcastle on the 30th August. This was the second Bishops war.

Charles was strongly advised to negotiate a truce, and summon Parliament yet again. At the treaty of Ripon (1640), it was conceded that the Scots were to occupy Northumberland and Durham and be paid £850 per day for their quarter; they were also to be reimbursed for their expenses involved in going to war. In the following negotiations, agreement was soon reached, Charles making many concessions and paying £300,000 to the Scots who returned to their side of the border. Charles was under considerable pressure in many directions; he made no attempt to intervene when Laud and Stafford were impeached and sentenced to death, civil unrest was on the increase. Charles really needed to unite the various factions in England, a similar problem faced by his mother in Scotland earlier. Neither managed it and it is probable that Charles did not even try. He viewed the Puritans and the Parliamentarians as opponents to his authority and set about to destroy them – his normal modus operandi. A split occurred in Parliament and in the Country; there were riots in London against the King. Charles ordered the Attorney General to indict 5 M. P. s and one member of the House of Lords, who he believed were responsible for inciting the London riots and who had previously encouraged the Covenanters to invade England. A herald went to Parliament

---

1 The river Tweed was the border between Scotland and England

requesting that the five members be handed over, the House refused, citing Parliamentary privilege. The next day Charles with a body of soldiers went to the House of Commons entered and demanded the five members give themselves up, they were not there and he was effectively hounded out. The fall-out from this debacle, damaged the King irreparably, he and his family fled London fearing for their safety and went to Hampton Court. Parliament controlled London, civil war had been likely for some time; it was now inevitable. Parliamentarians and Royalists attempted to build their positions and in August 1642 Charles raised his standard at Nottingham Castle, a call to arms, and in effect a declaration of war.

All parts of the Kingdom – village, town, city, split. Generally the upper class were for the King, the lower for Parliament, but there were many other factors that came into play, mercenaries and conscripts. A major division was religion: - Puritans for Parliament; more conservative Protestants and some Catholics for King. Cornwall and Wales for King. Initially Scotland was for Parliament, but that went against the Parliamentarians claim for Englishness, the situation changing in 1645. The King drafted in soldiers from anywhere, something that eventually drove many to change sides.

The combination of the Covenanter army and the well trained and organised Parliament New Model Army was far too strong for the Royalist forces, although they did have some notable victories. Eventually, (April 1646), a defeated Charles fled from his Headquarters in Oxford and surrendered to the Scottish Army, who held him in Newcastle upon Tyne. Naturally this was not a situation that the Parliamentarians were happy about, and Charles attempted to exploit the friction that this generated between the two previous allies. Parliament drew up a basis for a treaty with the King called the Newcastle Propositions. Although Charles was in no position to decline the offer, he was never likely to accept, and 'played for time' which he used to try to enlist assistance from European Royals. Both the Scots and the Parliamentarians lost patience and when Charles tried to escape, the Scots were reimbursed 'for their troubles', handed the King over and returned yet again to Scotland. Charles was taken to Hampton Court Palace.

But Charles was not finished: he modified his stance on the Newcastle Propositions and wooed the Presbyterians in Scotland and England. It was this new alliance of Scottish, Presbyterian and Royalist interest, that enabled Charles to escape his rather loose captivity in Hampton Court Palace, and start the second civil war of 1648, in the firm belief that he could now win, regain his throne and power. His judgement in this was no better than in many others made during his reign; defeat and capture was swift.

Oliver Cromwell had become an MP in 1640, whist still in the Army, and he continued to operate in both political and military arenas. He was one of those who in 1647/8 attempted to negotiate some form of settlement with Charles, but found his intransigence impossible to deal with and when Charles escaped from his Army captivity, to start the second civil war, Cromwell set about the King's downfall with a vengeance and was much involved in bringing him to trial.

By early 1649 Parliament had been purged of all Presbyterian sympathisers, resulting in what became known as the 'rump Parliament'. Many of the remaining MPs encouraged by Oliver Cromwell, believed that as long as Charles was alive there would always be a risk of further unrest, he did, after all, have form. So in January 1649 Parliament charged Charles with treason against the people of England. He was tried, made little effort to defend himself, found guilty and beheaded on 30th January. Europeans and many English people were shocked; they could not believe that England had killed its King.

By 1651 all Royalist resistance had been crushed, Cromwell was made Lord General and after his attempt to create a 'Parliament of Saints' failed, he took complete control becoming Lord Protector of the Commonwealth of England, Scotland and Ireland. Relative religious freedom was introduced except for Quakers, Catholics and Creeds. But Cromwell who was a Puritan, was very strict regarding religious observances, and banned many 'pleasure' activities. Gradually the population tired of the strict rule and began to hate him. Oliver Cromwell died in 1658, his son Richard took over but did not have the same enthusiasm as his father and in 1660 stepped down; Charles II was asked to return from exile and become King. Restitution and severe retribution followed, but for most of the population the natural order had been restored.

It is against this background that the story of this Northumberland beekeeper and his treatise is set.

Chapter 3

Northumberland Beekeeping and the Civil War

On April 9th 1603, James VI of Scotland arrived in Newcastle upon Tyne on his way from Scotland to take possession of the crown of England following the death of Elizabeth I, to become James I and thus unify the crowns of England and Scotland. He passed through again in 1617 on route to and from Scotland, the only other time he left England during his reign, both visits reinforced the Royalist ties with the town. Knighthoods were conferred on each occasion. In the intervening period (1609/10), Newcastle experienced its worst outbreak of the plague/pestilence up to that date.

In 1624, Newcastle resident X inherited a 'small fortune', which included a house one mile from Newcastle and a property in the Northumbrian countryside. His good fortune allowed him to indulge in different activities away from his trade, and he declared that his intention was to turn his attention to "countrye affairs", with honey-bees being the top of his list. He bought one stock of honey bees in a straw skep[2] and kept them at his residence in Newcastle. He also purchased a hackle to help with weatherproofing the hive, despite his later stated dislike of them. This was the standard 'kit' for keeping honey bees at the time and it would have been unlikely that there was any alternative for him to purchase. A later entry indicates that he paid 20/- for the package[3].

Skep.

Skep with hackle.

---

2   Skep is the Anglo-Saxon word for basket.
3   Equivalent to about £150 today.

The skep and the style of beekeeping associated with it had remained almost unchanged for many preceding centuries, the knowledge being passed from generation to generation and although X was literate there was little published which taught an alternative. In England the skep hive was usually constructed from straw, although heather and willow were also used. A continuous straw rope was assembled, generally with the aid of a truncated cow's horn, and then 'sewn' into a spherical shape – close to the natural shape of the cluster of a colony of honeybees. A small piece was removed from the bottom ring to form the entrance. It was strong, well-insulated, and inexpensive. Most agricultural labourers would have been able to make one, and could easily get the raw materials. They were made in varying sizes but three pecks; three quarters of a bushel, was considered the standard at this time[4]. Writers often defined the size of a skep by the number of rings it comprised: this is of course, not accurate because there was no standard for the diameter of the rope or the skep.

Once made it was considered necessary to 'dress' the skep. The loose ends of the straw on the inside were trimmed, and then the inner surface would be smeared with a mixture to make it more acceptable to the bees and thereby increasing the chances of them remaining when introduced into it. There was no standard for this, but a mix, generally of herbs and strong beer or honey was common. Some beekeepers of the day enlisted the assistance of a pig to lick out the deposited mixture as the beekeeper rotated the skep, with the pigs head inside. It was claimed that the froth from the animal's mouth added to the attraction for the bees. Thankfully, most writers also suggested wiping around afterwards with a linen cloth, the skep that is. Two cross sticks were wedged in the skep at right angles to each other, about a third of the way from the top prior to the bees taking up residence. This helped support the combs and stop the skep collapsing inwards. X wrote: -

….take two handfuls of pease malt, barley &c (but malt is best) and put it into the hive……and let a hog sow or pig eat it all out still turning the hive as hee eats that the froath may remaine of all sides; then wipe the hive with a clean cloth……..the sticking of it with quartered hasell (*hazel*) or sallow (*goat willow, pussy willow*) sticks; this sticking of a skep is only to keep the top of it from falling down and the sides from bending inward when its laden with honey…..

At other places in the manuscript for small sticks he uses the terms 'spelkes', 'saver'. Spelke is local dialect and still in use today when it is often used to describe a splinter, especially when it has inserted itself into part of a small child's body. 'Saver' is described by X as a small slip or small stave, but I was unable to link this to present day Geordie.

---

[4]   As a comparison the British National Brood box of today is approximately one bushel.

The skep would then be 'cloomed' in an attempt to make them weatherproof. A mix was prepared which generally contained short straw, sand or ashes and cow dung, which was plastered over the outside of the skep. X used "yellow clay and the hardest cow dung … fresher made the better … half of each, let both be very cleane….". Further weather protection was provided by a hackle – a further layer of straw, sometimes cloomed, in the shape of a cone that fitted over the skep and used in winter. Some beekeepers used wooden boards for cover but pans and the like were not thought good. The skeps would be supported on planks or individual stools, bricks etc. to keep them off the ground.

The whole system was based upon the bees natural propensity to swarm when they were housed in such relatively small hives. Having brought colonies through the winter, they would be watched during spring swarming times; the swarms were retrieved using techniques similar to those of today. A linen sheet was spread upon the ground and the swarm shaken upon it; a skep was placed close to the heap of bees with one side supported on a saver. The bees gradually occupied the skep. Near dusk the skep and its bees were removed and set upon a stool or plank next to the parent colony. The skep was cloomed and sealed to its support. It was usual to try to limit the number of swarms from each hive to two, the prime swarm and one cast. To stop further casts, beekeepers provided extra space under the hive by propping it up: bricks were commonly used although the straw nadir did exist. X however, appears to have used a 'dake'[5] – a square, hexagonal or octagonal box, 3 inches to 4 inches high, open top and bottom, with an entrance for the bees.

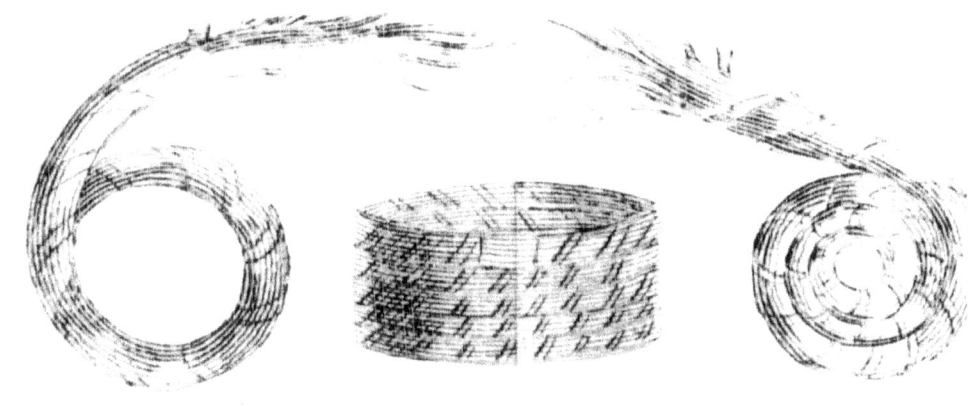

Straw nadir and construction

---

5   I was unable to find any other reference to this term, and it is not part of present day Geordie.

To dissuade a swarm from returning to its initial alighting point, X recommended brushing it with nettle or hemlock having observed that bees did not settle on either of these.

In his detailing of hiving swarms X said it was wrong to go "muffled up" because if a bee stings the glove all the bees go for you, picking up on the 'rank smell'. He advised: -

...If you must have anything on your hands or face let it be your knit gloves of thread and a network scarfe of silk or linen over your face and neck, for they cannot endure woollen or leather soe as if you had a shirt or such like about you it would please them better than any other apparall.....

At the end of the season, colonies were assessed for the amount of honey and population of bees present. To determine the amount of honey present hives would be 'hefted', (lifted on one side), and to assess the number of bees in the hive, the beekeeper would tap the skep put his ear to the hive and listen, the longer the buzz the greater the number of bees present. Our author's technique was to blow in the entrance and then placing ear also at entrance to determine the length and level of the noise. It was the same technique that he described when assessing a colony for purchase: -

...now when you buy to know the best you must set your mouth to the passage where they enter and blow into it as hard as you can, then presently lay your ear to the place where you did blow in, if you hear a great and long continued sound or buzzing of bees be sure there is good store of them,.........

I suggest giving this latter method a miss at home!

X also had something to say about the colour and temperament of bees when purchasing them: -

...know the quietest and best bees by their colour for they are be browne and smoothe then they are right but they that are black and rough are fierce and not soe good ;

Thus reinforcing the belief by many that the old English bee was brown not black.

Most authors then spent many pages describing how to determine which colonies to retrain for the following year. Basically: - those heavy and with many bees were kept, the others destroyed; the aim was to go into the winter with the same number of stocks as the previous year. Destruction of the bees was achieved using the brimstone pit: rags would be soaked in brimstone and set alight, placed in a small pit and the hive put over it until all the bees were dead. The combs were then cut out of the skep and the honey cropped.

The system was very cruel; but was beekeeping knowledge and skill such that an alternative was possible? The answer from this manuscript is clearly yes. Open and closed driving were known and practised by some, but it is doubtful that the average farm labourer would have used it. However, X questioned the benefit of driving, believing Charles Butler's view that bees only lived fourteen months[6], so there was little gain in trying to save them. He also wrote of driving: -

.... One reason for not burning as some people would have it is merely pity for it is unreasonable say they that wee should destroy that poor creature that has laboured for us, but this is easily answered by reason that all things were made for the use of man; again say they a hive that is driven may live; but this is but a hazard for scarce one in five doe....

So X had used driving, but his driven colonies rarely survived. At this time distance it is difficult to understand why driving would be responsible for the death of a colony. The more likely explanation might be associated with the fact that it was it was the weaker colonies that were the candidates for driving, and driving in this instance was to save the bees not the colony.

Closed driving

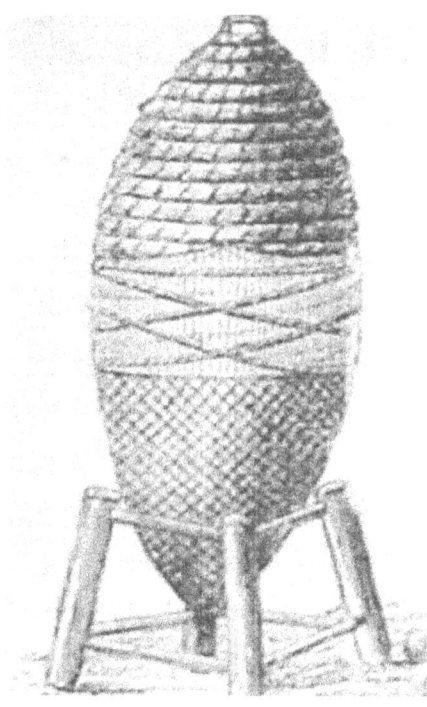

Closed driving

The hive to be taken was turned upside down, positioned in a frame, (X used 'dake' again here) or bucket. The empty hive to receive the bees was placed over the occupied skep and a cloth wrapped around the join; the bottom skep was then rhythmically tapped for about fifteen minutes. The bees ascended into the upper hive, which allowed the beekeeper unhindered access to the honey and use the bees as he pleased. The technique was also used to unite colonies but it was acknowledged that union would not be achieved without some fighting amongst the bees; however, beekeepers of the day believed that the moment one of the queens was killed the fighting would stop. This is closed driving.

---

6   Rev. Charles Butler, *The Feminine Monarchy*, 1609. Unusually Butler's logic was flawed in this instance.

Open driving

Open driving

The top hive was held at an angle using 'driving' irons or two pieces of wood. The bees were watched as they ascended and the queen plucked out as she travelled over the join between the two skeps. Open driving could also be used for uniting colonies, but because theoretically, the queen would be one of the first to move up, and removed, much of the carnage was avoided. This is open driving and gives greater flexibility than closed. Later in the manuscript X wrote that Dr Wilkins, Master of Wadham, Oxford, at the time, had told him that: -

...the beemaster at Oxford undertaketh at any tyme to show the queen in the hottest summer by driving one hive into another and then with his hands bare in the middle of company search the queen bee and discover her while the bees fly about his ears and hee not stung by them... and Dr Wilkins who told me he was an eyewitness and sayeth that for his remedy he doth place a smoking brand somewhere in the grasse when the wind brings the smoke upon the bees and that he washes his hands and face with wort[7].

What is true today, also applied in the 17th century. Increasing the number of honey-bee colonies can only be achieved at the expense of a reduction in the honey crop. X concentrated his efforts on expanding his apiary[8]. There was a change of Monarch in 1625, and in the same year Newcastle was further visited by the plague. However, by 1627, X had increased his one stock to thirty, without purchasing further colonies. More than half of his stocks were sat on two 'spruce deals' at his Newcastle residence. These colonies were overthrown, crushed and destroyed, which prompted him to move the

---

7   Wort is the liquid extracted from the mashing process during the brewing of beer.
8   James Bonner in his book *A New Plan...*, 1795, wrote extensively on this.

Driving irons

remainder to his house "in the fields" one mile from Newcastle. However, here they were neglected, did not thrive; were attacked by vermin, particularly mice and wasted by degrees until rogues came along cut out the crowns of the surviving hives and stole the honey. Eventually all X's bees died out. Up to this time X had housed his bees only in skeps.

The national problems referred to in the opening chapter were replicated in all cities, towns etc. and Newcastle was no exception. It was a Royalist city, but in 1632 there was a riot by apprentices because a new kiln and ballast heap had been made without the gate of the town called sand gate. On June 3rd 1633, Charles I stopped off on his way to Scotland for purposes explained earlier. On the 4th he was entertained by the Lord Mayor to dinner, whom he duly knighted. On the 5th he visited Tinmouth Castle[9] and we are also informed that some of his party played 'goff' on the Shield Field. He stopped overnight on the return journey also. All intended to bolster the Royalist support in the city prior to the inevitable civil war.

The Plague revisited Newcastle with a vengeance in 1636; in 8 months about 5,000 residents died, nearly half of the population. Fumigation by pitch, resin and frankincense was used to try to stop it spreading. Streets became covered with grass due to lack of trade and use. All those who could, fled the City, amongst them X who went with his family to his residence in Northumberland. By 1638 whilst still residing in Northumberland, X started to keep bees again. Initially he walled in a small piece of land to accommodate his bees and built a small house therein to house one of his servants. The wall was "fower[10] yards high to keep out the rogues"[11]. He bought 'fower' hives and moved them onto the site in winter. Initially they were also much harmed by mice, but with care thrived. Because of the damage by mice X resolved to put his swarms into hives made of wood. During this year X appears to have travelled to Lincolnshire, where he met "one that claimed to be a reverend divine" who told him a story that poor people gathered honey dew in dishes "and made use of it for honey", illustrating how readily available it was in that County. Also in 1638 magistrates of Newcastle

---

9   Tynemouth Castle and Priory are situated on the headland to the North of the mouth of the Tyne and was therefore, strategically important.
10 Local dialect for four.
11 The English/Scottish border regions had been subjected to lawlessness for many centuries by gangs known as the 'Border Rievers'. Logically the hostilities should have ceased when James I came to the throne, but personal and inter-family feuds continued long after. Hence the need for security for his bees.

raised £600 by means of a land tax to provide military stores for safety and defence of the city; civil war now a certainty.

On 5th May 1639 Charles I entered Newcastle at the head of his army and all peers, on a march against the Scottish Covenanters. They stayed for twelve days; the Mayor and town clerk were knighted. Charles moved on to Berwick where he negotiated a settlement with the Scots, after which he returned South again through Newcastle. It was a 'phoney' agreement, and as reported in the first chapter the Scots were back and in control of Newcastle in 1640, it was held by a garrison of 2,000 until early August 1641 when they accepted a financial settlement and returned to their own side of the border.

By the late spring of 1641, X had increased his honey-bee stocks to over 100, and when the King yet again passed through Newcastle on his way to Scotland this time to seek an alliance with the Covenanters, many of his party visited X to see his bees, verifying the number and the splendid condition that they were in. The only record of this is from X himself. The King returned to Newcastle, on the way from Scotland on the 19th November having been unsuccessful in his quest. By February 1642, X had lost many of his colonies in what he called "these woeful tymes", but he still had fifty remaining.

Having declined the advances of Charles, the Scottish Covenanters entered into alliance with the Parliamentarians and on January 15th, 1644, their army again crossed the England/Scotland border, this time at Coldstream on the way to take the strategic city of Newcastle upon Tyne. They again travelled through Northumberland unopposed and on the way destroyed the remaining hives of X[12], who almost certainly was not at home at the time. By February Newcastle was closely invested, but the outer wall was still secure. The Scots were in no hurry, and when the Northern Royalist Army was defeated at the battle of Marston Moor in July the outcome in Newcastle was inevitable. The outer wall was eventually stormed, surrender having been declined. It was soon breached in several places and the Royalist principals and their friends shut themselves in the castle. Lesley, Earl of Leven, leader of the Covenanter Army, was not prepared to accept anything but unconditional surrender and on the 22nd October the remaining Royalists marched out of the castle and surrendered. Newcastle was now held by the Parliamentarians. On the 19th November 1644 the House of Commons passed an order that the following 'Delinquents and Malignants' be sent to London: -

Sir Thomas Liddell, Baronet   Major Ralph Bowes
Sir Nicholas Cole, Knight and Baronet   Captain Cuthbert Carre
Sir Thomas Riddell Jnr,   Captain Edward Scott

---

12   This is consistent with the record in *Records of the Committees for the compounding etc. with delinquents and Royalists in Durham and Newcastle*. A Surtees Society publication, vol. CXI.

Sir Francis Bowes, Knight	Captain Thomas Blenkeship
Sir Alexander Davison	Captain Metcalf Rippon
Sir George Baker, Knight and Recorder	Captain Henry Marley
Sir Francis Liddell	Captain John Marley, Mayor
Mr Ralph Coale, Alderman	William Marley
Mr Ralph Cock, Alderman	Henry Rowcastle
James Coale, Sheriff	William Robson
Doctor Ambros	Captain Thomas Sharper
Mr Yeldred Alncy	Captain Thomas Pawle
Doctor George Wisheart	Sir William Riddell
Captain George Cock	Sir John Marley
Sir Thomas Riddell	

X, our delinquent, was imprisoned in Winchester House[13] London, where "having much idle time …… I thought it not amiss to employ some part thereof for the good of my country" – he wrote *A Treatise on Bees*. He claimed that he had written it before his confinement and it would be a repeat, as best that he could manage from memory.[14]

A fellow prisoner at this time was Sir Kenelme Digby, English courtier, diplomat and natural philosopher. His father was executed in 1606 for his part in the gunpowder plot. Kenelme was a founder member of the Royal Society and best known by beekeepers for *The Closet of the Eminently Learned Sir Kenelme Digby, Knight, Opened;* published from his notes by a close servant in 1669. It is known for its recipes for mead.

All those in office and rebellious against Parliament, were stripped of office and a committee set up to determine the Sequestering (seizure of property) of the estates of delinquents. X was among those affected, but unlike some he did not appear to actively negotiate a settlement with Parliament – payment of a 'fine' normally resulted in some or all of the seized property being returned. It might have been that he did not have the necessary finance. I was not able to find any Parliamentary records relating to any settlement.

---

13   Winchester House had been converted into a Prison by order of the House of Lords in 1642.
14   If this is the case there could be a second version in existence.

The Earliest Record of Beekeeping in Northern England

Chapter 4

A Treatise of Bees

Probably due to the circumstances under which the manuscript was produced, it lacks structure and there is little logic in the arrangement of the material. X tends to meander along often repeating himself. In style it is very much more of a manual than most other bee publications of the period - it is his way of keeping bees in his hive, with some natural history inserted as it came into his mind. The following is my re-ordering of his material.

<u>The Natural History of the Honey-bee</u>
Much of this material X obtained from *The Feminine Monarchie,* by Rev. Charles Butler, then vicar of Wooton St. Lawrence, first published in 1609; X had read the 1623 edition, and occasionally one suspects that he actually had it with him whilst in prison. He had also read and refers to *The Ordering of Bees* by John Levett; *A Treatise Concerning the Right Use and Ordering of Bees* by Edmund Southerne; most of the works by the classical writers, Virgil, Aristotle, Pliny etc. He demonstrates his ability to read and write Latin. Unfortunately he does not appear to have encountered *A Discourse or History of Bees,* by Richard Remnant published in 1637, the most advanced of books on bees, in English, at that time which might have led to him question some of his writings.

<u>The Queen</u>
Although the ancient Greeks were not convinced of the sex of the 'big bee' Western authors universally referred to it as King, for most, especially the clergy, it was inconceivable to think that a female could be in charge. However, Luis Mendez de Torres in his book *Tractado Breve de la Cultivacion y Cura de las Colmenas,* published in 1586 wrote: -

…The bee, called the mistress, without coupling with a male and without the pain of childbirth, produces a seed from which are engendered three kinds of bee – mistresses, drones and ordinary bees – according to the different cells in which the seed is placed…

Whilst not altogether correct, the central premise is accurate.

In his book of 1609, Butler stated unequivocally that the large bee was female, and to his satisfaction, at least, verified it. However, Butler also wrote that there were five orders of bees and X concurred. In addition to the queen, drones and common[15] bees, they added two more layers of command under the queen. X wrote: -

...the commanders have difference of ornaments for although non be comparable to the queen for stately shape and beauty yet if you will observe them in the spring and beginning of summer till near August you shall find as they goe and come in troopes (as Butler observed) with fine coloured sprigs on their heads like a feather some fower some three others two and one and very many with spots on their foreheads which are very likely to be badges of honour and command or else why should not every one be soe adorned for they have the same means to be so.

X often returned to a topic and reinforced his thinking on it. In this instance writing: -

...they will have but one supreme commander, although they have many inferior governors and others of good account amongst them as is evident by their ****** and other circumstances for there is in every small stock fower or five palaces for their queens lodgings and pleasures which are commonly made at the corner of a comb some about the middle others near the bottom and the top of the hive there in fashion and bigness resemble the boule (bowl) of a tobacco pipe full and larger in the middle as she may have roome to turne in and much narrower at the entrance where she goes in and out made thick and strong of the purest wax, when you find any of these stopped closed at the entrance be sure there is either the queen or some of her princesses in it imbalmed in it as I have often made tryall and shown to be divers for when you find any such you may split it open fully with your knife and find her within wrapped curiosly and close in the thin shining matter being clear and tough which being opened you shall see under her head a shining substance about the bigness of a wheat corne like yellow amber, but not soe hard and it hath odoriferous smell not unlike it when it burneth. the commanders have difference of ornaments for although non be comparable to the queen for stately shape and beauty yet if you will observe them in the spring and beginning of summer till near August you shall find as they goe and come in troopes (as Butler observed) with fine coloured sprigs on their heads like a feather some fower some three others two and one and very many with spots on their foreheads which are very likely to be badges of honour and command...

So although the queen was head, they considered that she could not do it without help which was, of course, male. Both Butler, X, and others claimed that they had seen these intermediaries, but it had much to do with the belief that bee society must reflect that of humans, than genuine visual proof. It was often

---

15   'Worker' was not used at this time.

recommended that when inspecting a colony of bees, a strong beer should be rubbed into the hands and face and some should also be drunk to disguise the breath. Could this explain the impaired vision?

Although X accepted that the single large bee was female, he illustrated his unease with this by quoting the following non p. c. ditty initially written by Francis Quarles in the 17th century: -

Ill thrives that hapless family that shows
A cocke that's silent, and a hen that crowes
I know not which live more unnaturall lives,
Obeying husbands, or commanding wives.

Most authors were able to accurately describe the form of the different orders of bees; X was no exception other than he had five to describe rather than three, but he acknowledged that he was not aware of their precise roles. X claimed to have read Aristotle who provided some basis when he wrote: -

There is a division of work in the hive: some make wax, some make honey, some make bee-bread, some shape and mould combs… others smooth and arrange combs.

X also recognised that if the queen dies "the colony never thrives thereafter" and states that he had seen her feed herself with her fangs. Of the queen's colour he wrote: -

…all the master bees that ever I saw, differ little in colour from the other bees, but that their legs are yellow inclining to a golden colour…

The Bee[16]

Rev. Charles Butler was of the opinion that bees "live not above one year and two months". Whilst his reasoning was somewhat illogical, and at one point he confused the queen and the common bee in his explanation, X accepted this and consequently concluded that there was, therefore, little point in trying to avoid killing them to get the honey, reinforcing in his somewhat cold blooded approach quoted earlier.

X believed that bees had the same senses as humans – sight, hearing, smell, touch, and taste: he gave sound reasoning, as he saw it, for each. He knew that they did not "defile their hive with their excrements", easily deduced from seeing them fly whenever possible in winter, but rather bizarrely declared that "Every hole (cell) being six square according to the number of the bees feet": he was not alone in this thought process.

---

16   Worker.

The Drone

X knew that the drones were male, and that they were bred in the same manner as common bees. This he deuced from noticing the larger cells on the combs. He also believed that the drones contributed to the work by being the water carriers, adding: -

...Bees need water to mix with the coarse stuff they fed the young with for they use much water in mixing the course stuffe they gather to feed their young ones and drones, which stuffe they bring on their thighs there are some flowers that when the dew is gone relays a fragrant dust as the great gooseling and the palm that grows upon the sallows and the like when it is soe dry that they cannot make it stick on their thighs then they rowle themselves in it till they fill all the hayers of their bodyes with it and so bring it home they carry all liquid things .... in their bags...

As other beekeepers, X knew that the bees ridded the hive of drones in autumn, he had observed it, and he derived that this was a means of saving the food that they would otherwise consume during the winter. He acknowledged that some beekeepers aided the process but he did "not like a sparke leap or engine rods"[17] approach. He mused 'why drones?" –

...bout the latter end of July or the beginning of August they drive and kill the drones (which are the male bees) there is none of them left being soe as if there were noe use of them else nature would not be soe prodigal as to breed such creatures which doe nothing but devour much honey and drink much water...

So he clearly felt drones had some use in the grand scale of things but was struggling to determine what that was. However, he was convinced that drones had some role in the reproductive process, saying that if drones were not present there would be "no breeding happening", adding -

....soe heavy and full they can hardly either fly or goe making a hard shift to fetch their water and when they have drunke they come home to feed again - drones some (*conjecture*) they are bees that have lost their stings but such dye presently leaving the gut behind with the sting; but to confute that opinion and to prove that they are bred as other bees are besides the reasons before given, if you take but notice you shall find certayne combs in every hive much bigger than those the bees are bred in, in which they are bred...

---

17   Disappointingly there was no explanation of this statement.

Reproduction

Whilst seventeenth century beekeepers believed that both the queen and common bees to be female and drones to be male, they had no knowledge of how they 'engendered'. It had not been witnessed mainly because beekeepers did not know where to look or what to look for. This resulted in many hypotheses. X regurgitated most of the possibilities without offering a specific opinion: -

(i). Drone and common bee engender by copulation, assumed inside the hive, it would after all, be unseemly anywhere else: it was necessary for each new bee.
(ii). Gather young ones upon flowers.
(iii). Queen engenders all.
(iv). Queen and bees lay eggs then sit on them for 48 days, result bee.
(v). Blow' in the cells as fly or wasp, egg often referred to as 'bee spat'. Fly spat is still used today in relation to flies.
(vi). Reproduce like fish.

The clergy in particular preferred to believe that the drone played no part, thus eliminating copulation and many thought that queens propagated queens, common bees propagated common bees, and drones propagated drones. X believed that drones played some part:-

...Its likely therefore that the female doth make of them till they have conceived and then they send them packing; in what manner they conceive is uncertayne for they are seldom or never seen to *engender*

Had X encountered Remnants book of 1637 he would have read that the queens: -

...have neer their stings a little neat place for receipt for generation ...... there is in the hinder part of the male or drone a little white thing like the instrument of generation; take one of them alive and crush the body of it somewhat hard between your fingers, and you shall see it put forth.

How Bees Gather Honey

Beekeepers of the period had no idea of the relationship between the bee and flower; X believed that the bee took little from the flower otherwise it would die: -

...I am fully persuaded they gather little honey upon flowers but almost altogether wax....

Elsewhere he wrote: -

The bees take little from the flowers or herbs but a small tincture and the quintessence of the ordinary dew which falleth on them mightily...

...The northern bees are forced to gather most of their honey from the flowers and herbs the virtues whereof it retayneth and they will never labour on pure and liquid dew in the flowers so long as they can have any of the clammy honey dew which they bring to honey with much more ease.....

Basically X and most others believed that honey came from honey dew and bee food from pollen, which he termed course stuff, or ambrosia. The former view was reinforced by his conversation with the Rev. Divine in Norfolk. Notwithstanding he gave details of the flowers etc. that the bees worked and it is interesting to note that he knew buckwheat to be useful for bees: -

.....they love Buckwheat towards the latter end of July and all August when the heather or heath is in the flower they will quite give over the meadows and almost all other flowers and labour altogether there on that, for it gives great store of liquid honey which they bring home in their little honey bags soe that in the six weeks or thereabouts that it lasts they furnish themselves better than all the year before.....

Victorian beekeepers were known to plant Buckwheat for their bees and it will be raised again later. However, our author believed his bees gathered most from 'herbs and heath'.

Swarming

Beekeepers were well aware that bees swarmed, knew how to catch and hive them afterwards; it was a major part of their management system.
Butler stated: -

...a swarm doth consist of all such parts as the stock doth; namely of a queen-bee, Honie bees as well as old and young, and drone bees...

He reasoned that without some of each component part of the colony it could not prosper, adding that it was the common bees that decided when a colony would swarm, bee population being the over-riding consideration and weather permitting. He added: -

...hive bee full, so that it may be divided at the least into two or three sufficient companies; one to remain with Marpefia the old queen, another to go forth with Antiope the Prince, and a third Halpie, which together with unripe brood in the celles may take another swarm to serve Orithya...[18]

This might be interpreted that he artificially swarmed stocks, but I believe he is referring to the bees dividing themselves.

---

18   This was Butler's quaint way of putting it: in Ancient Greek and Roman legendary history, Marpesia was the queen of the Amazons; Antiope was her daughter who succeeded her after she was killed in battle; Orityia was another daughter. It is interesting to note that Butler got the sexes and relationships correct, but referred to Antiope as 'Prince'!

X did not fully concur with Butler on this matter – he believed that it was mainly young bees that left with the young queen. It is difficult to understand how they got this wrong because when they hived the swarm they were able to view the bees entering: did they not observe carefully, or not look at all? X wrote of the indications to be looked for after the issuing of the prime swarm: -

…when you see the drones come abroad after they have cast their first you may look for the second swarme within eight or ten days and so of the rest if you listen at the hive at night for you may hear the young lady begging leave of the old queen (for soe Mr Butler calls for) to goe out with her young army, bothe whose voices you may distinctly hear, first the young are calling toute, toute &c divers tymes over with a treble and shrill voice the other answers in the same notes and language with base and lowere voice; which latter if you hear then the old queen gives her consent to the young ladys request and then be sure they will swarme the next day if the weather be seasonal…

As he indicates, this is exactly as Butler describes it. This belief led to major discrepancies in their thought process regarding developing a logical management system. They did not believe that it mattered where the swarm was placed when caught; X placed it next to the parent hive and allowed nature to take its course in sorting out which hive the bees returned to after foraging. However, X did acknowledge that bees were less likely to sting when in a swarm, and: -

…if any of them chance to use their stings, but gently with noe intention to harm…

If only.

Chapter 5

Development

Although skep beekeeping would continue in the same form for a further three centuries, X had several issues with it. The first of these that he attempted to remedy was the damage that his hives and hackles suffered from mice and other vermin who viewed it as a suitable winter residence: it was warm and dry, and cloming only provided a limited barrier to access. However, X believed the skep was an ideal home for his 'mother' colonies, so he sought a means of providing a better defence against the vermin. He first added a small square 'dake' 3 inches high of a size to fit snugly around the bottom of the skep. A piece five inches long and three-quarters of an inch high was removed from the side to coincide with the skep entrance. The gap between the skep and the dake was filled with clome, which was allowed to set. A wooden pyramid hip roof of a size to fit over the skep and rest upon the dake was constructed, providing a far superior cover than the hackle, and given a sufficient overhang it would shed water from both hive and stool. The roof and dake were joined by nailing, insulation in the form of duck down or short straw could be added between the wooden roof and skep. This he believed would provide an impenetrable barrier to the vermin and weather.

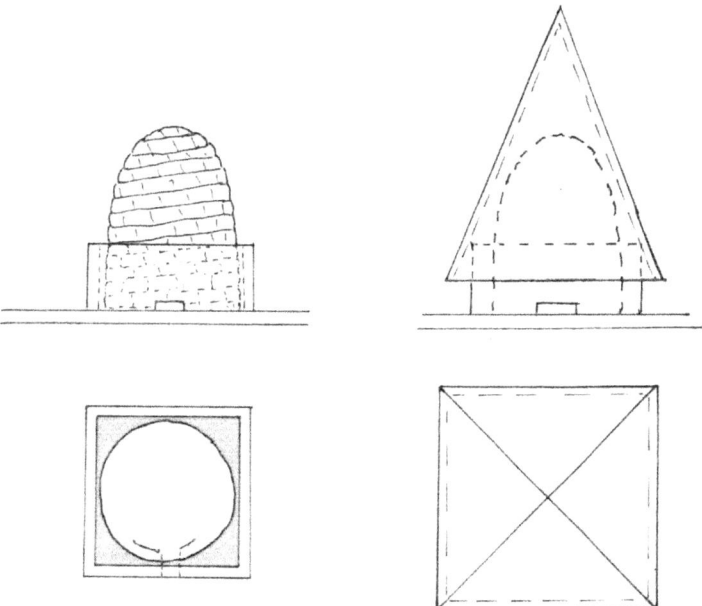

Skep with dake and pyramid hip roof.

Unwittingly, X had devised what was probably the first genuine double walled hive.[19] As well as outwitting the mice it had the advantage that the added insulation lead to better heat retention in the hive and therefore increased the chance of swarming and brought it forward. He mused in the manuscript, that this arrangement would be better if the dake to skep connection was not so permanent and it might be better with insulation under the roof that could be discarded when it was removed for the summer; nailing the two parts together would have been an inconvenience. However, even in this raw state there was nothing comparable in existence.

X did not let it rest there. He reasoned that the skep served no useful function as a home for a swarm, which he viewed as a honey-producing unit. He had also witnessed two swarms that had risen approximately at the same time amicably joining together where they pitched, but of course he was not certain of what he was seeing. However, he reasoned that he could build large honey producing units by putting many swarms together. The belief prevalent at the time was that the carnage when uniting two colonies would cease when one queen was dead, but he also suggested that sprinkling both colonies with strong beer before putting them together would lessen the problem. Unfortunately he did not give details of how he added a fresh swarm to the existing conglomerate.

For his honey gathering hive X initially tried a box which had one side loosely attached by straps and clasps but found that it was too troublesome, disturbed the bees and "spoyled their work". It was also size inflexible, and did not lend itself to easy management. However, he acknowledged that with this arrangement it was possible to view the bees working writing: -

...for soe you may sit upon the whole side and see them unladen themselves and doe their business within and never trouble them much better than in any glass hive as some dream of...

Despite this advantage he abandoned the idea, and for his next attempt: -

...I had also another fashion of hives made of beach made round all hoopes near the fashion of a large bouy or marker for ships...

These held approximately 30 gallons, equivalent to about three BS brood boxes, into which he put three "great swarms". However, he found that the sun made the hoops give way, starting at the bottom, leading to a spreading of the vertical pieces, allowing vermin and robbers access so he abandoned the idea.

---

19   Much earlier there had been hives housed in outer casings, which might be considered 'double walled'. In a report on the inspection of Hertfordshire Association member's bees in the late 1800s, it was stated that one member cottager had adapted his skeps for the collection of honey by adding sectional supers above the skep with the crown removed to provide bee access. The skeps were encased in a rough square box with chaff filling the gap between the two; to this a roof was added. To add the section crate the roof was removed, crate placed on top of box and the roof replaced.

Development

Perhaps different material for the hoops would have solved the problem, but this appears not to have occurred to him, so we are not informed as to how he intended expanding the hive.

X does not inform as to the length of any trial that he gave of either of these two hives but the indication is that it was very limited because he quickly decided that the answer was to have a hive of many boxes giving flexibility, especially in regard to management, but had he persevered with either of the first two he would have achieved that aim with those also. He initially trialled two boxes but found three "to do the turne". The size of the complete hive was to be 13 inches square and 30 inches high for first swarms and 11 inches square and 26 inches high for second swarms (casts). The boxes to be of Firr deals[20] "for they delight much in that wood but ash or oak will suffice". All wood needed to be well seasoned, which he interpreted as laid up for one year. Front and back were to be the thickness of "whole firre deals", and the sides to be half that thickness. The sides in the top two boxes were made shorter than the front and backs by one inch in order to accommodate shuts or shutters. The size for the larger of the two boxes approximates to that of a skep, and two thirds that of the British Standard brood box. It is unfortunate that he does not give accurate details in regard to the shutters and their operation.

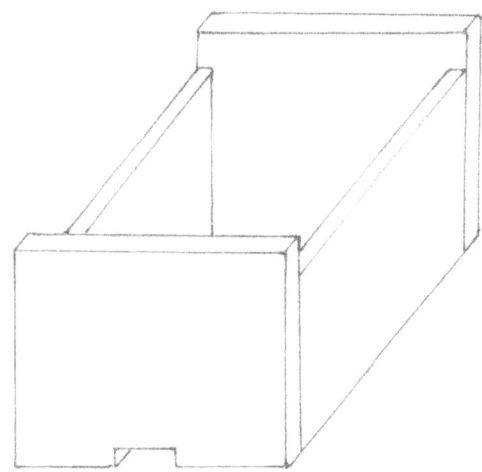

Box with short sides

The roof was nailed on to the top box and of a size to give three inches overhang on the sides and two inches front and back to shed the rain. Initially only the bottom most box was equipped with access, but he reasoned that by providing

---
20   See Appendix III.

the same facility on each box he would reduce the work load of the bees because when returning home laden, they would not have to ascend through the combs with their load. It also aided interchangeability of the boxes. Then: -

….there must be a ledge made of deal five or six inches long and two broad placed on the top of the two lower chambers for the bees to light on when they come home laden soe they may enter at the passage that are in the chambers above the great passages in the bottom of them; I mean you may put two wood screws in these ledges and have two holes bored in the hive and soe fasten them…

Adding: -

…many small holes forced with a piercer so large as that two bees may passe by one another as you find convenient five seven or more as you please…….

Observing: -

…wherein they tooke such delight and content both in the ease they found by the quick dispatch of burdens as also for aire in hot and close weather as you may well perceive by their numbers they continually labour and sport about all parts of the hive and you shall seldom see them make use of the lower passages for labor until they have filled their hive down near to them as they fill their hive with honey they will winter stop most of their small passages about with tough matter somewhat like shoemakers wax for it hath an excellent smell like the gum that is in the firre woode…

He suggested that if 'firre deals' of this size were not available the boxes could be modified in shape but should be of similar capacity. Whatever the shape the boxes needed to be clasped together to prevent movement in wind; he proposed "two pins on each side and bound fast with jack thread" - an early hive fastener still appearing in 19th century books. He recommended that the bottom box be given a coat of straw in winter, constructed by joining together a series of small bundles to create a blanket that is wrapped around the box[21]. The entrance has to be kept clear. A similar cover could be provided for the roof, the mice apparently no trouble now. He also provided an additional cover of "tough shar fired earth thin pared".

---

21   This is similar to that used by some beekeepers of the period for the construction of hackles.

Development

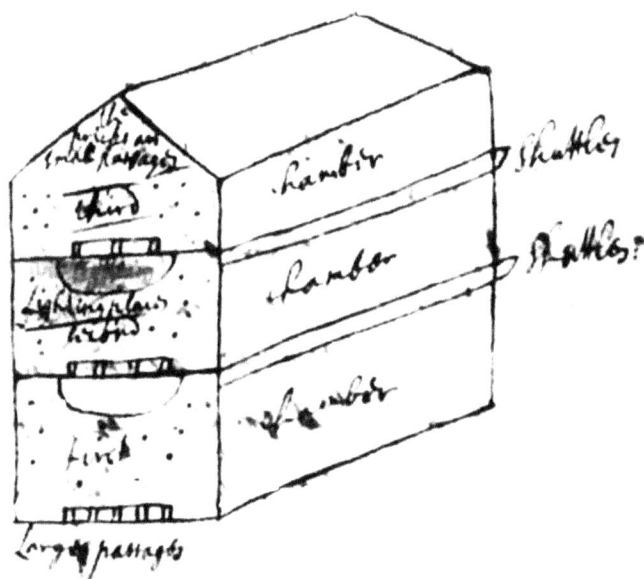

Box hive of three boxes from manuscript

He gives his modus operandi thus: -

Year One
The prime swarm was shaken into the top box which was then placed on a stool. If the weather was inclement for a few days X put a comb of honey from another hive into the box, fourteen days later another box was added under them and when they had filled that one, a third was added under that. The judgement of when to add the next box was by hefting and listening at the hive. To stop the parent colony from swarming a third time, a 3 inches dake was added under the parent hive, size matching that fixed to the skep.

Unfortunately his descriptions are sometimes confused, as in this instance. The implication from his basic description of the hive was that the roof was permanently attached to the top box[22]; this would make it non interchangeable with the other boxes. This would also make it difficult to shake the swarm in as he describes – box would need to be placed over swarm that had been shaken out onto a sheet, or inverted onto the roof to receive the swarm. Also, in his first description of the hive he did not include entrances in the top two boxes making the system of management described impossible: this gives the impression that it was something that 'developed' as he wrote.

The boxes were dressed and spelked as with the skep, but there is no mention of a participating hog.

22   At one point in the manuscript he states 'nailed on'.

The Earliest Record of Beekeeping in Northern England

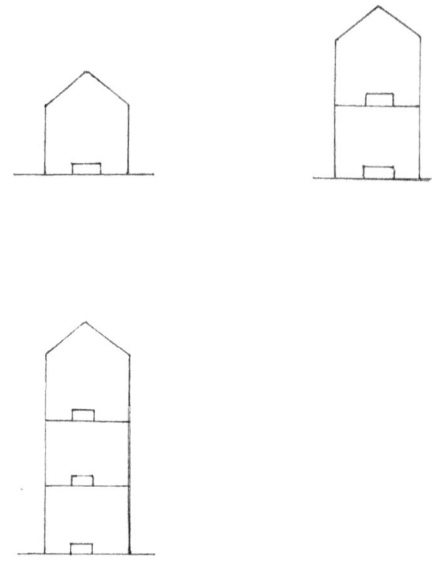

Year one operational diagrams

Next X gave specific details of how the hives should be arranged in the apiary. The strongest swarms were put in rows next to the old stock, the second swarms were set in rows below these, but the reasoning leaves a lot to be desired!

At the end of September each of the box hives were assessed for honey content, those found light were fed. X described the method of feeding he utilised when using skeps – open feeding. At sundown those hives not requiring feeding were blocked up. The next day at sunrise, combs witch contain honey in open cells were placed on planks in or near the apiary and then simply "whistle out your bees". Surprisingly X had missed the possibility of feeding at the base of the hive via a drawer of some description, or just a shallow pan, easily provided for in a square box hive.

In October covers were put on hives, if there was the likelihood of these being attacked by mice the straw covers were coated with "glassye clome" – clome mixed as before but with crushed glass added to the mix.

At the end of October, all passages in the top boxes of the weakest hives were stopped up, and the entrance restricted. When it snowed all passages were stopped up leaving some small holes for the bees to breath at; once snow had gone the passages were opened for the bees to fly and excrete. He related an instance of a hive swarming very late in the year and settling in a neighbouring hive; when inspected X found that there was no food in their original hive – a starvation swarm?

Year Two
In Mid-February, the hive was lifted and the stool cleaned off using a long knife.

If weak hives were seen to be the subject of robbing, X utilised a home-made funnel constructed from thin planed board. It was 6 inches broad x 7 inches long x 1 inch high; internal measurements. He first stopped up the entrance of the hive being robbed, brushed away the robbers and inserted the narrow end (one bee size, internally), of the funnel into the hive entrance. The robbing bees inside the hive, exited, those outside could not enter because the legitimate residents were capable of defending the small entrance. Or just simply reduce the size of the entrance! A search of my 19th and 20th century catalogues unsurprisingly did not reveal one merchant offering such a piece of equipment.

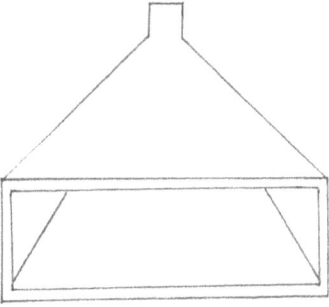

Anti-robbing funnel

As the spring progressed the entrance size was increased, (funnel had presumably been removed when robbing ceased the previous year), and at the end of May the winter coverings were removed. Leaving the insulating coverings on until this time would encourage breeding and swarming, although he intended these hives to be primarily for honey production.

About August 1st the honey in the top box was taken. Two nights earlier, the hive (all three boxes), was lifted and a 4 inches dake placed underneath it, and all entrances in the top box closed; this operation could be improved by initially inserting two 'savors or stays' under the hive a few hours earlier. According to the plan the bees moved down vacating the top box. A long knife was inserted and cut through the comb between the top and middle box: the shutter(s) were then inserted isolating the top box. The top box was smartly removed to a bee proof location and an empty dressed box with roof and its entrance open was put in its place. The shutters or 'shutt' was then removed; enabling the bees to re- occupy the top box and two nights later the dake was removed.

X made a point of saying that he used a sword rather than a knife to cut through the comb between boxes, advising that it was better done by two people, and that the blade end would require a cover! He does not, however, give details about how, when not being used, the hole in the top of the boxes should be sealed; it is difficult to see how removal of whatever he used for this purpose would be an easy matter, the bees having attached comb to and propolised it. X added that using two shutters, one from each side would be better than a single shutter, and that the operation would be aided by nailing a couple of spelkes at the top and to the insides of the box, as a guide for the sword and shutters. Unfortunately he did not provide precise details which might indicate that he did not believe that there was any originality in what he was conveying.

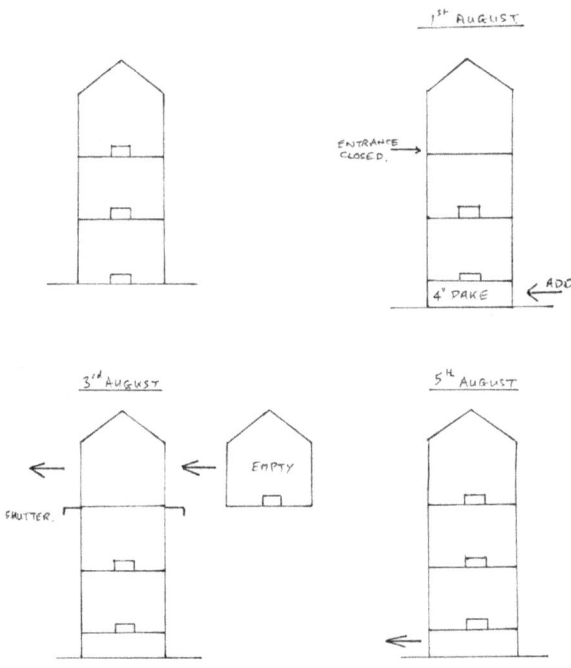

Operational diagrams year two

Year Three.
Other than the usual business at swarming time the hive was left until early November when the bees were in the top two boxes. X cut and shuttered between the bottom two boxes: the bottom box was removed, and the remaining two boxes were placed on the stool and shutters removed. The honey and wax in bottom box was harvested whilst the colony was left to winter in the two remaining boxes. It was left to the beekeepers assessment if he considered it possible in early August to also remove and replace the top box and its honey as previously.

Development

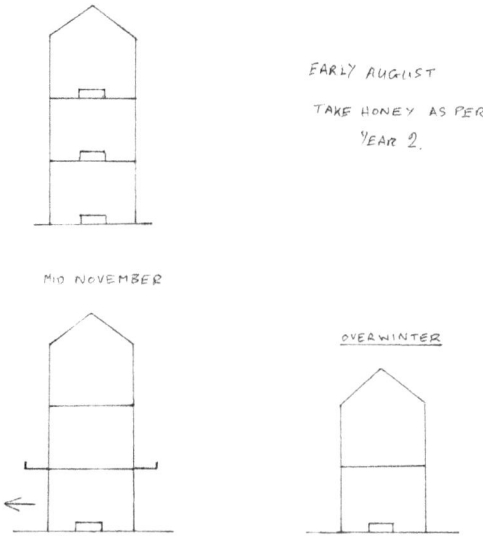

Operational diagrams year three

Year Four

In spring, the colony was allowed to swarm; the swarm was collected and hived in another box. The bottom box was then replaced. Depending upon the season it might have been possible to take the honey in the top box as previously. In early morning, mid-November, the top two boxes were knifed and shuttered; a dake was next inserted above the shutter and the shutter removed allowing the bees to rise. Two days later the shutter was re-inserted between the top of middle box and the dake; the bottom two boxes were then cut and shuttered. Next the middle box was removed and its honey and wax harvested. The operation on the dake for the middle box had to be reversed either at this stage, or early spring the following year.

The Earliest Record of Beekeeping in Northern England

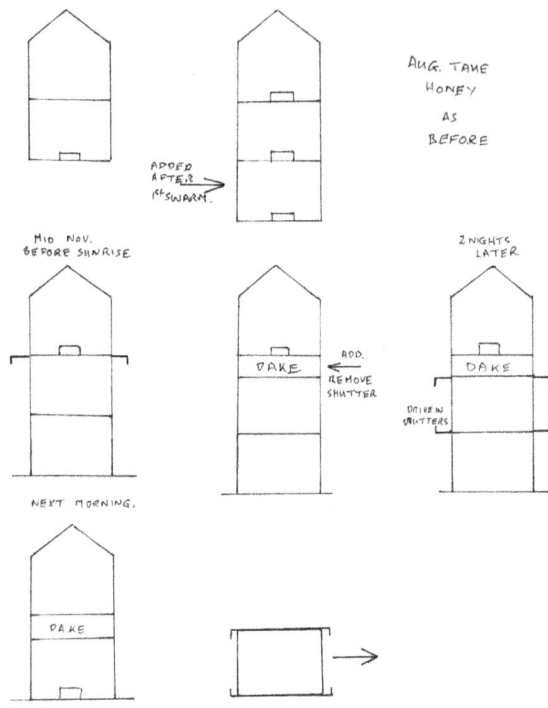

Operational diagrams year four

Year Five

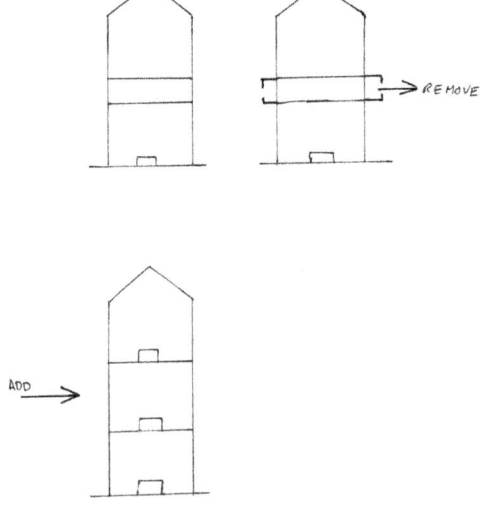

Operational diagrams year five

The cycle of operations was then repeated. X summed this part up thus: -

...thus have you the way made playne and open by which you may be merciful to this poor diligent creature keeping them in health to labour for you continually with little care and trouble by taking the wax and other filthy slopping from them ...Which they cannot thrive nor prosper that they may have liborly to renew their wax you take which keeps them sweet and in continuall hive. This is the mistory of ridding the bees from filthy wax and slopping ; not found out or practised before for ought I ever read of or could heare which they prevent naturally of themselves when they have rooms and convenience provided; but yet I doubt much (not withstanding this playne demonstration of the inconveniences and the neydes for the preserving of themselves) that the sloath and negligence of many will hardly suffer them to follow my advice in this way; and therefore I will for the good of themselves perswade them to spare killing of their bees at least till such tyme as they have a competent number.....

It is highly probable that at least the final year of the five year cycle described had, not been trialled, because he hardly had the time.

But X was not finished and stated that there was a third way "which is the perfection of this mistory". In May before the large stocks (the honey producers), swarmed, which he tested by heat and noise, as detailed earlier; a larger chamber was added underneath the stock as shown.

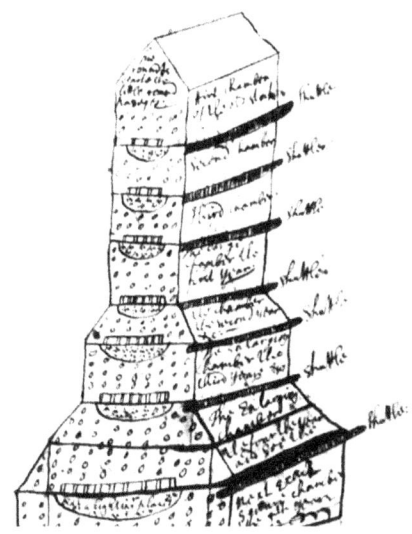

Large honey producing hive.

Being two foot square, this box was of larger cross section than the hive that it was being added to and therefore a slanting roof was provided to join the two and shed the rain. Because of the size, the shutters on top of this box needed to be in two parts, meeting in the middle. The purpose was to stop the bees swarming by provision of space. Late August the top two boxes were taken away and replaced with one for the duration of the winter and clomed up. The following year an even larger box was added to the bottom etc. He is working on the principle that if the bees are prevented from swarming the colony will grow ever larger. Swarms were continually added to aid the process, and he contends that 40 swarms in one hive was good adding that as the expansion continued they might need props to keep them up!

If this appears somewhat fanciful, compare it with the pyramid hive of Rene Antoine Ferchault de Reaumur[23] in his *Memoires pour server A L'historie des Insects,* vol. V, 1740, and his development of it shown here.

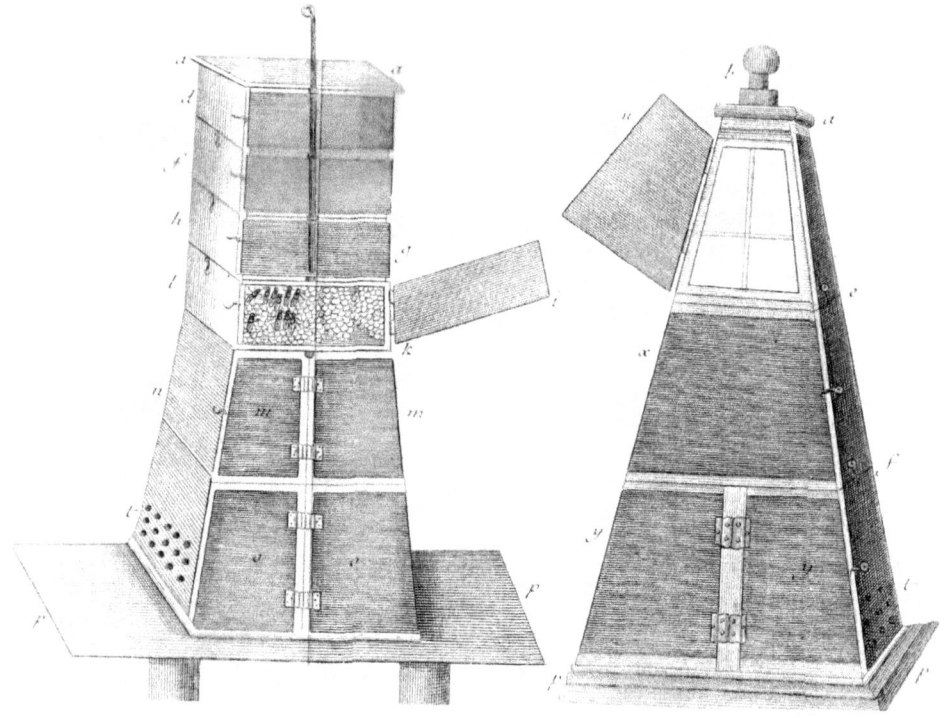

Reaumur's pyramid hive

Whilst X proposed these very large hives he recognised the problems that this caused so when one of the large hives became very unwieldy he proposed splitting it. To achieve this he suggested placing an empty top chamber behind (the East side) the large hive, distanced by about eight inches and connected to the large hive by a six-inch square 'truncke'. It was necessary to have corresponding pieces

---

23   Reaumur was a French scientist, known for his work in many fields; the six volumes on Insects was considered the definitive work on the subject at that time.

removed in both boxes so that bees could travel from one to another through the corridor. The young bees were supposed to move into the empty box, but he acknowledged that this might not happen. If so, most entrances to the main hive were blocked up to persuade them. When the new box was well stocked with bees, the tunnel was removed and the hole at the back of the old hive sealed. The new colony was placed on a new stool, and then simply "hold your course on taking the profit of the old stock".

X proposed an alternative method: - make a large chamber as big as the bottom two chambers of the large hive, omitting two deals on the back to coincide with the lower two entrances on the occupied hive, and place it in front of the large hive. Bees have to travel through empty box, and younger bees will build comb and establish in it. As before it is removed and placed on its own stool in the apiary, the two missing deals having been 'softly' replaced. However, he accepted that the larger hive and splitting of it, was not from experience – it was simply conjecture - again. Realising that this left him open to criticism, he wrote: -

It is very likely these things will be slighted as a meer conceit to say the truth it is no more yet it is proved and hath great reason for it considering the nature and proporty of the subject and without doubt all the rare things in the world were noe more at first; and it may be many of them scoffed at till they did vindicate themselves by experience which hath been the mind of many brave designs……..

The significance of the operation just described will not have escaped the beekeeping historian; it is a form of collateral hiving and very similar to that described by another beekeeper from the North-East of England, Matthew Pile, in his book *The Bee Cultivator's Assistant,* almost two centuries later. Pile used skeps.

X proposed other possible solutions to the imagined problem:
(i).   Force them to swarm in early season. This of course, gives the same net result as his first two ideas
(ii).   Make larger rectangular hives, which would require an internal framework to support the combs – spelkes of three inches diameter were mentioned. Similar internal frameworks were incorporated by Rusden and Gedde in their hives a century later.
(iii). Add rings onto the sides of the boxes and make a hoist, the 'patterne' of which he illustrated. Three to four men were required as was a cart equipped with a wooden vessel to catch the runny honey on it, and a horse to pull it. The boxes are lifted and swung around onto the cart.

The Earliest Record of Beekeeping in Northern England

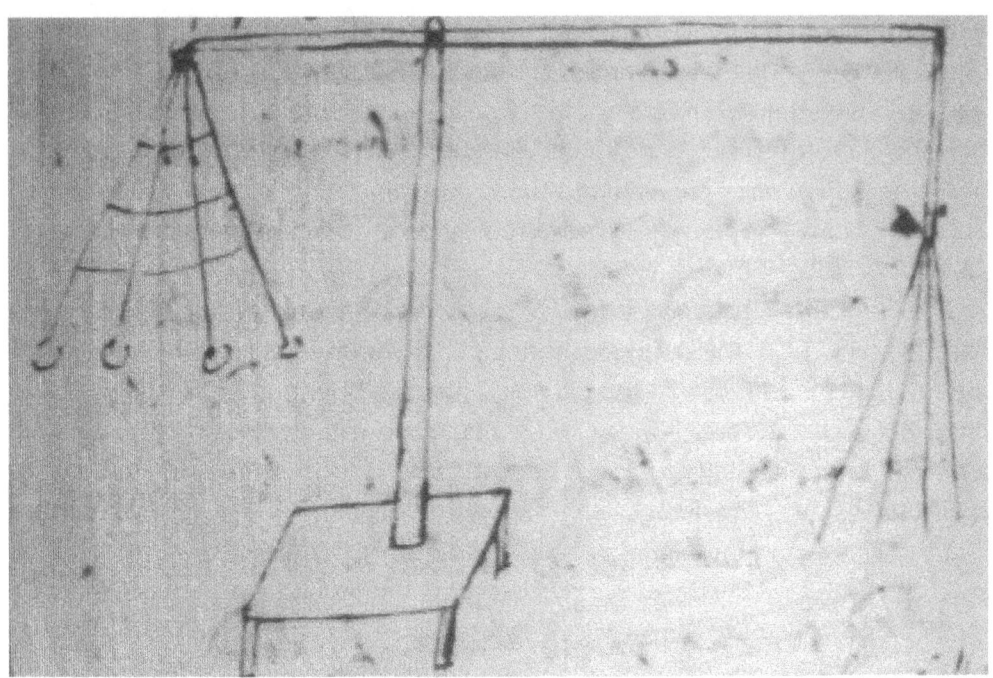

Method for lifting very heavy hives

X specified the amount of honey required for a colony to over winter successfully – one and a half full boxes, unfortunately he did not inform as to which of the boxes he was referring to. However he acknowledged that it was necessary to get an accurate measurement of how much honey was in the boxes to be retained for winter. This was done by use your wimble, which needed to be at least 1 ½ feet long and "webbed from end to end". My interpretation of this is that this is similar to taking a sample out of a cheese or apple corer.

He concludes his treatise with a brief description of how to order honey. He strained honey through a 'haire cloth' made into a sack 6 foot long and 3 foot across, with a hoop at the mouth and another 2 foot from the bottom, hung from a frame of rafters in his house with a 'vessall' placed underneath to catch honey. X stated that the purest honey should be separated and done on its own. He either had or wanted to convey the impression that he had large crops of honey.

Wax was boiled with water, scum removed and left to cool. Wax removed when solidified. If 'cleane' wax was required, he poured the boiling water and liquid wax through a suspended Socrates bag[24] of course hair, collecting the water and wax in a bowl underneath.

Following his release from Winchester gaol, X appended a very detailed table of contents. This was a repeat of the 'headings' that he had initially included in

---

24   Socrates bag – conical in shape as sleeves Socrates used on his shirts. Still found in most active kitchens.

the margins of the treatise manuscript. At certain points in this list of contents he included additional material which he had obtained since his release from prison. These give valuable biographical information in addition to bee craft information.

X first relates that at some time after his release he met with Dr. Wilkins, who at the time was the master of Wadham College Oxford. This has to be after his release from Winchester gaol in 1649 because Wilkins was not Master of Wadham until 1648. X wrote: -

If you are inclined to make a very fine present of honey combs observe but this story which was related by Dr Wilkins that ingeneous person and mathematician who told me that the Beemaster at Oxford when hee would provide a dish of honey combs for a present; take an empty hive in the *head* wherof he placeth a pewter dish which was held up with cross *beams* then into that hive he puts a latter swarme; where they will worke and fill the dish with fine white combs and honey; a present for a prince…

I have already given his quote re driving/bee-master of Oxford, and thirdly: -

…they gather much upon the wheat blossom as appears by a remarkable story which Dr Wilkins relates of the beemaster at Oxford, whose practice is, when he hath a weake stock to carry it into a wheat field at the tyme of blossoming and there to leave it for a fortnight by which tyme they will be well stored with honey and much refreshed…….

It is probable that given the story above he is referring to Buckwheat[25], no longer grown in the UK. There is some circumstantial evidence that 'the bee-master of Oxford' was Rev. Clark, rector of Drayton, Oxfordshire, a hamlet of 17 adults, 2 houses, Manor house and rectory, at the time; the living was in the gift of Anthony Cope, Baronet, who appears to have had an interest in bees and was an acquaintance of Wilkins.

X bought bees again in 1654 and was still alive in 1658 when he reported that he was operating his large stalls again, but he gave no details of the hives he was working.

Sometime after 1655 he visited Norfolk, where he met a Mr Clark, whom he described as a "gentleman of credit". There is no evidence that this is the same Clark as above.

After his return to Northumberland, X borrowed a copy of *The Ordering of Bees,* by John Levett, and proceeded to make notes as he read: these are appended to his treatise and form the latter part of the manuscript. His notes are under the same headings as Levett, which gives organisation to this section of the manuscript, sadly lacking in X's own part. However, he claimed to have read it before his incarceration and there is considerable evidence in the early part of his manuscript to reinforce that opinion.

25 Buckwheat is not related to wheat. It benefits from pollination, the honey from it being dark-coloured.

The Earliest Record of Beekeeping in Northern England

Chapter 6

The place of 'A treatise of Bees'

So, is there anything original in this manuscript that informs about beekeeping in addition to what has traditionally been considered as the avant-garde of the craft in the 17th century? Historians have generally used a book by Samuel Hartlib, *The Reformed Commonwealth of Bees*, published in 1655 as the basis of information on beekeeping during this period with Charles Butler's earlier publication of 1609 as the authority on bees and beekeeping using the skep hive.

Samuel Hartlib has been variously described as polymath, educator and intelligencer. He was born in Prussia about 1600 to an affluent family[26], relocated to England in 1628 to avoid war. In England he gained patrons, opened a school and developed a circle of contacts and correspondents. During the English civil wars he published works mainly on agriculture and whilst he avoided contentious areas, there could be little doubt as to which cause he supported, and he was eventually awarded an annual pension by Cromwell of £300. His vision was to have centres of knowledge in every reasonably populated town, serviced from a central hub - basically him and his 'inner circle'. This arrangement was labelled by Hartlib and his fellow collaborator John Dury, (a Scottish Calvanist minister), 'The Office of Public Address' which they envisaged would also act as a labour exchange putting the poor – especially the intellectual and religious, in touch with benefactors. Upon the re-establishment of the old order in 1660, Hartlib lost his pension and died a pauper two years later. When the Royal Society was formed in 1660, gaining its Royal Charter in 1662, Hartlib was not a member, although many of his 'circle' formed the early core membership. However, it is natural to draw parallels between the Royal Society and Hartlib's circle; many of the founder members of the former were also part of the latter. Hartlib had simply supported the wrong side.

The Reformed Commonwealth of Bees contained information that Hartlib had elicited from individuals by correspondence, mostly after 1650. Hartlib was not a beekeeper, but there are signs that he felt that he held a position of authority, and some of the information was not provided very willingly.

The first communication reported was with Mr Carew[27] of Antony, 'great

---

26 Samuel was the product of his father's third wife, who had two sisters one married to a Mr Clark son of a Lord Mayor and later to a very rich Knight, Sir Richard Smith a Kings privy councillor.
27 Probably Sir Thomas Carew, (Baronetcy purchased by his father), who was a regicide, and hung drawn and quartered in 1660.

husbandman of Cornwall' who revealed that he used a hogshead wine cask as a bee-hive, accessing the honey via a hole in the top; but no real management is described. He also believed in the rotting carcass hypothesis for the generation of bees, which is a good indicator of his level of knowledge.

There followed a discourse left by that "zealous publick-hearted and learned gentleman Thomas Brown[28], Dr in Divinity and of civill law, prior to his leaving for overseas". His method was based upon preserving the bees and thereby gaining colony size, believing that bees naturally increase in numbers of their own accord, not dissimilar to X's opinions. He refers to natural nests 6 to 8 feet long, which he did not believe could be the work of one swarm, and that bees did not desert their birth residence. He stated that bees work top down, and their hives should be same size all the way down, with flat tops and bottoms so that they can "easily set one upon another" and "because the round figure is the most perfect, I rather choose it", but he acknowledged that some made them square with four boards. Casks which had "Honey, Muskadine, Canarie or Malaga wine" he considered to be the best. Size – "just a bushel" within, breadth = 1 1/3 x height[29] AB in diag. D round hole 3 to 4 inches wide.

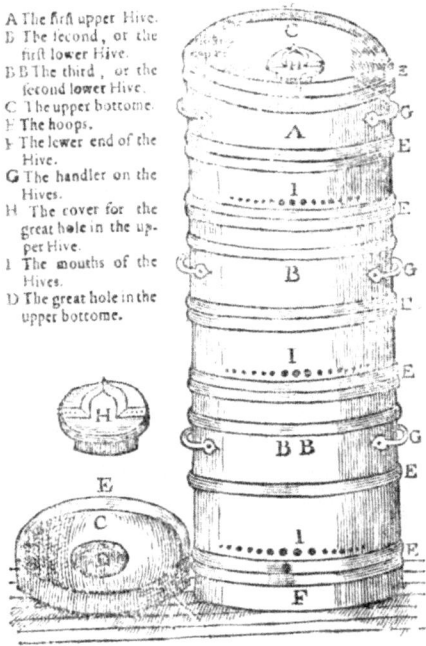

Dr Brown's cylindrical hive

---

28   The Hartlib papers show that this is Dr Thomas Brown who wrote from St James Parish in Barbados "I forgot to tell you that there are no bees here of no kinds". I believe that he was a student from Christ Church College, Oxford prior to emigrating.

29   Results in a height of approximately 16" and diameter 21"; just happened to be the size of the casks to hand.

In the diagram E are the broad hoops – he says six, 2 in middle, 2 at each end but the sketch shows seven. Bottom-most hoop large, 1 inch above the very bottom of the cask. There are basically 3 casks set on top of each other, A, B, BB. Both B and BB have what appears to be narrow hoops part way down. A hoop of iron or brass is added to F, giving greater strength, something that X missed in his assessment of this style of hive. Brown has three large holes to give access for the large bees (drones and queens?), and six smaller ones each side for the smaller bees. Uppermost hoop is 1 inch wider than the rest. He proposed an internal structure to support the combs, they resembled plant supports – rings with legs. This latter inclusion of internal supports pre-dated Rusden and Gedde but not X.

Brown started by putting bees, usually a swarm into A. When the bees had worked A down to its bottom, B was added under it with the interconnecting hole open. The bees eventually worked down leaving A and enabling the safe taking of it and removal of the honey. Further hives (boxes) were added as required. When the hive was 3 or 4 boxes high, a new prepared hive was placed as near the mouth of existing hive as possible, with its bottom slightly raised; bees will enter and at next swarm they will occupy it. This appears to have been an early swarm catcher but unfortunately there is no indication of its effectiveness.

There are some parallels between the methods of X and Brown, particularly the method of splitting colonies which X described in greater detail. Browns overall management system was simpler but it did not include information on how to deal with old comb. There is no date given for Brown's information and although he thought round hives to be superior, he acknowledged that the same system would work using square-boxes.

Hartlib commented that he thought square boxes would have been an improvement on the round casks, because a bill or drawer could be placed underneath for feeding. Surprisingly X also missed this continuing to use feeding in the open as he did for skeps. Initially this appears to indicate that whilst not a beekeeper, Hartlib was reasonably clued-up on the craft. However, the suggestion was more likely to have been made by one of his inner circle, who would have been given the Brown information to assess – Cheney Culpeper was the usual conduit for beekeeping matters.

There followed a letter which was a translation from High Dutch,[30] on how they do it "beyond the seas" but with no author credited.

This correspondent had witnessed hives with glass, covered with wood which had doors in[31] for viewing, but this only helped "considering of their nature". The

---

30 As spoken in the Netherlands at that time compared with that used in South Africa.
31 X tried a similar arrangement.

writer claimed that an experienced bee-master did not allow more than 2 swarms from a hive per year – they then raised the hive "by some inches", to give the bees more room to build down[32]. He also wrote that there were many attendants in a swarming hive but drones, who are fed by the labours of bees, are killed when bees have no further intention of swarming. He noted that young bees were idle just before swarming. He claimed that his method stopped the bees flying away at swarming time, and kept them from breeding drones and further generals[33]. He did not have to kill his bees to get the honey. In his accommodation (hive), bees went downwards when laden, up when un-laden, a common thought process in this period. So no swarming, no drones, no new queens, no killing, less effort: Beekeeper's Utopia.

His hives were horizontal, on two long poles kept in a garret close under the roof. The end of the hive that touched the roof tiles was 'closed' and an entrance hole cut in upper part. The wide end of hive stands was 'clapt' down on a plank, and was shut up with a bottom of straw pinned on all sides with wooden skewers. He used straw hoops to lengthen hive as he wished, effectively an eke used horizontally. Hives became 2 to 3 yards long.

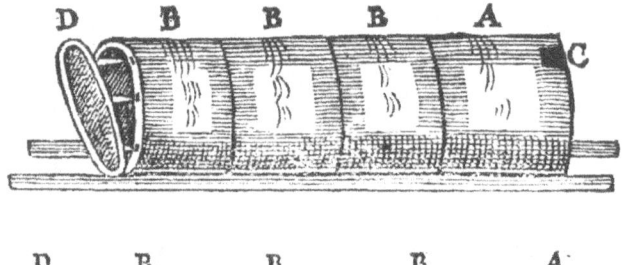

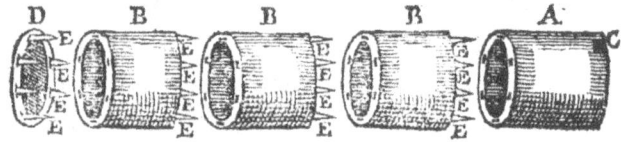

    A A common Bee-hive.
    B A prolonger to lengthen or eeke out the Hive withall.
    C A hole cut in the upper end of the Hive A.
    D A bottom or dore to shut up the Hive, whether it be single as A, or lengthened as A B B.
    E The wooden pins in B and D for the joyning of them to the ends of A or B.

Dutch recumbent hive

---

32   X also gave this method – it was clearly common practice.
33   Presumably these are the 'attendants' mentioned earlier in the paragraph

Honey was removed by unpinning the wide end and driving bees to the narrow end using smoke from burning linen rags. No illustration of the hive or meaningful dimensions were provided and there was no indication of the author or translator.[34] However he claimed that communicating this to Hartlib had cost him 200 + rix dollars[35], which reinforces the proposition in footnote 33. Another unnamed correspondent conveniently provided a diagram of a very similar, if not the same, hive.

Hartlib next dealt with the octagonal hive originally the design of the Rev. Mew. Because my research on this hive went beyond 1655 I have included it as Appendix V. It is sufficient to record here that there is no evidence that X could have been influenced in any way by the octagonal hive or anything written on it when he was devising his hive and the accompanying management strategy.

Following two contributions of little consequence Hartlib copies a section from *Lex Mercatoria* (The Ancient Law Merchant) by Gerard Maylnes[36], 1622, which inexplicably contains a few pages on beekeeping; the excuse seems to have been that Honey and wax are trade items handled by merchants. He gives details of a two part skep. This is the first record of such a hive that I am aware of. He detailed how to take the top part for honey: first raise the hive by several wreathes, bees descend, then draw a wire through between the skep and cap, and follow with a piece of parchment. This is similar to X with his dake, sword and shutters, but X did not list this publication in his treatise, so there is no evidence that he was aware of it, although it is my belief that X was a Newcastle merchant. Maylnes writes that bees are engendered using corrupted Heysar flesh in the same way that horse is used for wasps and man for lice. A very confused individual. He follows with a calendar of operations for the beekeeper. If he needed to feed in winter he turned the skep upside down and threw in a warm mix of wort, honey and water. To feed swarms he put dishes of the same mixture in the bottom of the skep, - refer Harlib's note on feeding with the square hive. If bees were diseased he set them over the smoke from a mix of cows or oxen dung, 'sophisticated' with sweet wort and the marrow of the ox or cow. The mixture needed to be well dried, apparently. Maylnes indicated that his reference was Hill's book of 1568.[37]

---

34 It is very likely that Cheney Culpeper brought this back from his time in Holland, but there is no information as to his linguistic abilities. Money flowed both ways in the Hartlib/Culpeper relationship depending upon the relative financial circumstances of them at any one time.

35 Silver coinage used throughout the European continent – an early euro! 1 rix dollar = 4/-.

36 Maylnes, English, b. Antwerp, became merchant in foreign trade upon return to England; failure of this business resulted in a spell in debtors prison which was somehow followed by success in government. At one stage he tried lead mining in Yorkshire and silver mining in Co. Durham.

37 Thomas Hill, *The profitable arte of gardening*. 1568.

Hartlib concluded with "Some writers of bees extant": -

Butler, *The feminine monarchy* 1634.

Henry Gurnay[38] – book promised but it never materialised.

John Levett, *The Ordering of Bees*, 1634.

Edmund Southern, *A treatise concerning the right use and ordering of bees*. 1593.

Richard Remnant, *A discourse or History of Bees.* 1639.

---

38   Gurnay; lived in Ellingham, Attleborough, Norfolk (St James) and West Barsham Fakenham, Norfolk, 1578 – 1648, (explains why book never appeared). Church of England clergyman. Could he have been the Rev. Divine that X mentions that he met in Lincoln? Is it the gent in Mew's letter to Hartlib? The Gurneys were a family of Norfolk Quakers, their name being associated with one of the four 'forms' of the religion. In 1683 a John Gurney was imprisoned in Norfolk gaol because he refused to take the oath of allegiance. Late 1600s he founded a Bank – Gurneys Bank.

Chapter 7
Time Line

1568	Thomas Hill, *The profitable arte of gardening.*
1593	Edmund Southerne, *A treatise concerning the right use and ordering of bees.*
1609	Rev. Charles Butler, *The feminine monarchie.*
1622	Gerard Maylnes, *Lex mercatonia.*
1624	X inherits "small fortune", bought skep and bees.
1625	Outbreak of plague in Newcastle.
1627	X has increased his stock of bees to 30 colonies.
1628?	Half of the 30 stocks destroyed – vandals, remainder moved 1 mile from Newcastle.
1628 -1629	X loses all his stocks of bees.
1633	Charles I visited Newcastle.
1634	John Levett, *The ordering of bees.*
1636	Plague in Newcastle. X and family move to house in Northumberland.
1637	Richard Remnant, *A discourse or historie of bees.*
1638	X buys four colonies of honey-bees in skeps.
	X travelled, met a "reverend divine" in Lincolnshire.
1639	Charles I visited Newcastle.
1640	Scottish covenanters in possession of Newcastle.
1641	Scots paid to leave Newcastle.
	X had 100+ colonies.
	Charles I in Newcastle again.
1642	X had lost half of his 100 colonies of bees by end of winter.
1643	Rev William Mewe summoned to the assembly of Divines at Westminster.
1638 – 1644	X carries out the research on his hive and management system.
1644	Scots retake Newcastle.
	X sent to London – Winchester House gaol.
1644 -1649	X, *A treatise of bees.*
1648	John Wilkins appointed Master of Wadham.
1649	X released from gaol.
	Assembly of Divines dissolved, Mew returned to his parish in Eastington.
1651	Samuel Hartlib, *Samuel Harlib his Legacie.*

1653	Wilkins and Wren put swarms into their Mew hive.
1654	X buys bees again.
1655	X visits Norfolk meets Mr Clark.
1658	X reports on his bees.
1650s	late. John Evelyn shows his manuscript *Elysium Brittanicum* to Hartlib.
1655	Samuel Hartlib, *The reformed commonwealth of bees.* John Evelyn, *Elysium Britannicum.*
1660.	Royal Society formed.
1662	Royal Society awarded Royal Charter.
1669	William Mew dies.
1675	John Gedde, *A new discovery of an excellent method of bee-houses and colonies.*
1676	Robert Plot, *The natural history of Oxfordshire.*
1679	Moses Rusden, *A new discovery of bees.*

Chapter 8

Concluding Observations and Unresolved Issues

I am in no doubt that X did not add to existing knowledge about the honeybee, his is a very practical treatise on beekeeping. However, there are signs that he attempted to verify some of Butler's hypotheses, but generally he simply agreed with them. His experimentation in this direction was limited to cutting open a queen cell and keeping grubs alive outside the hive in an attempt to study their development. The novelty in *A treatise on bees* lies with his hives and the management system that he developed to work them.

- X was operating a hive comprising three separate tiered boxes before Wilkins and Wren, albeit of a different shape. X had realised that it was necessary to start with swarm(s) in a single box and increase the number gradually, something which the Oxford group were still struggling with some fifteen years later. It is unclear how Mew operated his boxes; he declined to offer that information to Hartlib.

- When Hartlib pressed Mew for an answer on the financial return to be obtained from keeping bees, it would appear that he was unaware that Butler had already given a figure of £400 per portion for his gift of bees to his daughter and that Varro had written that 2 brothers were making £800 per annum. Later John Aubrey in his *Natural History of Wiltshire,* (1656 -1691) recorded that a Mr Harvey of Newcastle made £800 per annum from bees but there was no detail on the year that this applied to or the source for the information.

- It appears that the only octagonal hives that Mew had were those he described as a 'phancie' or 'Hieroglimph'. They were intended to illustrate to his parishioners the value of an ordered society and the benefit of work. His others were two storied but of what?

- When Hartlib pressed Mew for a description of how he operated his glass hive, Mew was eloquent about the ornaments on the hive but silent as to its operation. It was Gedde who eventually managed this.

- Many historians have linked Mew's hive with those of Rusden, Gedde, Thorley through to the Stewarton, but the only connection is the shape – they were all octagonal boxes. The same argument could be used between the hive of X and those of Bevan etc. right through to the British National

- and Langstroth hives. They are all square/rectangular boxes, but the same progression is never applied in this case.
- I believe that X's method to renew old comb and crop honey is original.
- X's 'swarm generating' hive was the earliest example of a genuine double walled hive. This discounts those hives that were given some form of outer layer – house etc. Mew's glass hive might well be included here, but there are no details available.
- X missed an obvious advantage of his square hive – the ability to add a drawer underneath to provide a means of feeding securely. He continued to feed in the open.
- Maylne mentions a straw hive with a super (cap). There were several who claimed to be the first with this in the 18th century. The skep and cap were separated using a wire, the honey then being taken from the cap. Although shown by Anon in *Traite des Mouches a miel,* Paris, 1690, the French did not like tiered straw or wooden boxes because they believed that cutting between the skep and cap interrupted the honey flow and led to bee deaths. This same criticism can be levelled at the hive of X. Gedde solved this problem by attaching a 'lid' to each box with a closable hole in, but this exacerbated the inter-box communication.
- X's method of persuading bees to populate a new hive appears to be original, but is only a variation upon the collateral principle.
- I have not found evidence that X was in Winchester house gaol with any other beekeepers, and it is unlikely that that his and Mew's paths would have crossed during this or any other period.
- X met Wilkins after his release from gaol; he might well have encountered Clark[39] of Banbury who was acquainted with Wilkins, at approximately the same time. Could this be the Clark who was the rector of Drayton, Oxfordshire, 1651 – 1688? Could Clark be the 'bee-master of Oxford'? He is mentioned in Plot's *The Natural History of Oxfordshire* as is William Tayler (a Northamptonshire man) who had moved from Oxford to Warkworth in Northumberland. Could Tayler and X have been known to each other?
- X also met a Mr Clark of Yarmouth, Norfolk and a 'reverend divine of Lincolnshire' at different times, but it is not possible to prove links between any of these.

---

39   A Mr Grant, in a letter to Hartlib, wrote of a Mr Clarke and apothecary of Oxford, who became the apothecary at the University of Dublin. "He hath much experimented about bees ……".

- It is unlikely that X met Evelyn, they were of such differing class, and X would have mentioned it – they were both royalists.

- The dimensions quoted for Evelyn's hive, from a sketch in his manuscript, could be an outer box, not the actual hive. Also of interest from the same drawing is what appears to be an outline of a pyramid hive; Reaumur 1734 had a pyramid hive and X's final imagined hive was of similar shape, all working on the same principle. But where did Evelyn get his from in the 1660s

- Hartlib had several correspondents on natural science matters but he mostly used John Beale (rector, Sock Dennis 1638, same, Yeovil 1660, Chaplin, Charles II 1665), Sir Cheney Culpeper, and Henry (Heinrich) Oldenburg. Oldenburgh was one of the first joint editors of the Royal Soc. Trans. and generally credited with introducing peer review. John Beale's wife painted a portrait of John Evelyn. A very close group.

- From the *Miscellany of the Abbotsford Club*, we are informed that John Gedde's wife lived in the North of England for a period including 1681; only some of this time was John Gedde away in London. Could Gedde and X have met?

- X mentioned an internal framework for his larger hive, but he had not trialled either hive of frame and gave little information on the internal support structure.

- A late 16th century book by Gallo[40], *Histoire de l'agriculture gauloise, gallo-romaine et medieval (History of Gallic agriculture Gallo-Roman and medieval)*, is said to mention stacked box bee-hives in addition to square horizontal versions. (Crane, *World History of Bee-keeping*/Kritsky, *The Quest for the Perfect Hive*). The horizontal hives were opened at the back and pieces of comb were removed for consumption, similar to early Egyptian clay pipe hives: there was no mention of sectionalisation. A knife was used to cut through the comb that was attached to the rear board. The illustration showing the upright stack comprising two boxes is just as likely to be of two separate hives stacked one on top of the other as one hive and with two boxes. A knife was again used to detach the top board from the comb and after the honey was taken the box was inverted to encourage the bees to replace the removed combs, which appears also to indicate only a single box.

- I am sure that my conclusion regarding the identity of X is sound even though it might be interpreted that the Hartlib papers create doubt, but this was due to his confusion with the information he received.

---

40   Gallo was from Bressica near Verona.

- Who is X? I believe the name given on the manuscript – *F Arm or Arm F* is not a name at all – it is the term applied to the son of an esquire or son of a gentleman, in the same way that Cler. Fil. is used to indicate the son of a clergyman. These terms were commonly used by Oxford and Cambridge Universities at the time to indicate the parentage of their entrants. I discounted Arm as an abbreviation for Armiger because it did not account for F and I was unable to link X with a Knight amongst the Newcastle Delinquents. I was also supplied with possible initials, but these do not appear in my copy of the manuscript and also proved a distraction. Following many hours searching my conclusion is that X was **Captain Edward Scott** and unless I find out otherwise I believe that his place in the country was at Bolam, Northumberland, but much of the present evidence for the latter is circumstantial.

- Captain Edward Scott lived and practised his beekeeping in Northumberland and his *A Treatise on Bees* proves that techniques in the craft were as advanced in at least one region outside the previously accepted centres for it, and possibly even ahead – whisper it! However, he devised a system of obtaining honey without destroying the bees, whilst having no problem with destroying bees in skeps to obtain their honey.

Appendix 1

Bonham catalogue entry

Bonham's catalogue for 28th September, 2004, contained details of a Letter from Wilkins to Evelyn. The catalogue entry read: -

Wilkins – Evelyn´s "dear and excellent friend" – was the principal founder of the Royal Society. At the time this letter was written, he was Warden of Wadham College, Oxford, and was to go on to be Master of Trinity College, Cambridge, and Bishop of Chester. Evelyn records in his diary for 13 July 1654: "we all dined with that most obliging, and universally curious, Dr Wilkins at Wadham. He was the first to show me the transparent apiaries, which he built like castles and palaces, and had so ordered them one upon another that he could take the honey without destroying the bees. These were adorned with a variety of dials, little statues, vanes, etc., which were very ornamental: and he was so abundantly civil at finding me pleased with them that he presented me with one of these hives that he had empty, and which I afterwards had in my garden at Sayes Court – where His Majesty came on purpose to see and contemplate it with much satisfaction" (Charles II´s visit being in 1663, Evelyn writing up his diary retrospectively from notes, although the entry above appears to contradict this).

Wikins letter read: -

"Honoured Sir. Your letter sent to Oxford, was returned back & found me here in London, wither I was by some occasion necessitated to come much sooner than I expected. I have here in readines for you, one part of the Bee-hive you desire, according to the same modell I have in Oxford. If you would desire to have two other like parts made to this (which I would advise) they may be done here in London by the same man who made this/I have taken order that it be left at Mr Bedles according to the direction of your letter where you ****** may call for it. I hope **** to waite upon you before my going out of town. And ***** always bee Sir most ready to serve you". Bearing Evelyn´s autograph docket ("From Dr Wilkin/Warden of Wadham Coll: Oxon: London 2: Apr: 1656.

This letter was purchased by the Beinecke Rare Book and Manuscript Library.

Appendix II

Some relevant extracts from the University of Sheffield's digitisation of Samuel Hartlib's papers

- "Mr Best hath invented a new sort of bee-hive." *Henry Best of Elmswell, Yorkshire, wrote his manuscript, 'Rural Economy in Yorkshire', in 1641. It does not contain any mention of a 'new hive' in his manuscript he used skeps. They might of course, not be the same Best.*
- Scot's of Newcastle. Date 1654. "Papers to Mr Rushworth or Executor Lampton concerning bees. 1654 - Mr Scots son or brother at Newcastle. All his observations concerning Bees or Bee-hives at least a certificate of his fathers or brothers yearly profit of 400 lbs. (*might be monetary pounds*), which he made of bees. 1654 – Newcastle dead man if any papers left. Remnants book of bees and another quoted by him. 1654. Mr Rushworth related of one Scot of Newcastle to bee a huge Bee-master undertaking to get 500 lb. a year by them. Hee lived to get 400 lb. a year but the scots coming to Newcastle destroyed or spoiled all his bee-hives. His son or Brother is yet alive who perhaps can give more information and a certificate at least of 400 lb. yearly gaine. Mr Rushworth." *This is the strongest evidence available that the author of the Yale manuscript is Scott rather than Blenkinsopp the only other real contender. However, the manuscript appears to provide information to the effect that X was not dead by this date and Remnants book was the one author that he failed to list or refer to. There appears to have been confusion, I think by Harlib between Scot and the Earl of Newcastle.*
- 1655. The questions that Hartlib put to Mew were fed to him by Sir Cheney Culpeper. In one communication on Mew and bees, Culpeper wrote "… wherein the whole ways of working of that little creature might be seene by which we might (I am confident) have sophisticated wines of our owne, cheaper and better than from other nations"; seeming to indicate that mead was imported.
- 1655. The chapter in *Lex Mercatoria* was informed to Hartlib by a Mr Brewerton.
- 1655. Letter Robert Wood to Hartlib. "Dr. Wilkins Bee-hive is accurately described in your treatise of bees, and for its transparency tis caused by a piece of glass a little bigger than my hand set into the hive on one side thereof." *This is the first indication of the size of the glass window, and verifies what I thought, that it was of little value. Robert Wood was a*

mathematician, printer of a Commonwealth newspaper, and one of the Hartlib inner circle.

- 1655. "I forgot to tell you there are no bees here of no kinds. Dr Thomas Brown from St James Parish in Barbados, April 13, 1654." *As near as is possible to get to show that this is the Thomas Brown of the cylinder hive. He was in communication with Hartlib, had left these shores, and was interested in bees.*
- 1654. "Enquire at Captain Mews his house at the upper end of Newgate Market towards Cheapside who can direct where Mr Mews a Minister in Gloucestershire lives who hath much studied the commonwealth of bees. Hee hath a whole amount of them in his garden and bee-hives of glasses. Mr Worthington." *Illustrates the difficulties of the time – there is someone I know of the same name (or close to it) so he must be able to tell me where his namesake lives. Also, Peter Mews, Royalist, who eventually became Bishop of Winchester, was the nephew of Thomas Winnliffe who was appointed Bishop of Lincoln in 1641. It is easy to become excited at this potential connection but X met his Rev. Divine in Lincs. in 1634! Mr Worthington was probably John Worthington, an academic, diarist, Master of Jesus College, Cambridge at this time, and communicator with Hartlib – part of the outer circle.*
- 1653. "Mr Angelo hath undertaken to write from Bristol to Mr Mew about the Commonwealth of bees."
- 1654. Sir Cheney Culpeper wrote to Hartlib with his take on Browns bee communication. It did not find its way into the book.
- N. D. "Mr Mews ladys recipe of mead like rock water."
- 1654. Richard Rawlinson (*Queens College, Oxford*) to Hartlib, 3rd Jan. "Mr Wren is to contact you re his bees experiment – expectations not fully answered. He did not visit Mr Benlowes before leaving London or he probably would have had his experiments."
- 1655. "The man of Gloucestershire the great experimentor of bees of a yellowish hair." Id Gous. *Did this refer to Mew or his bees?*
- 1659. Moses Wall to Hartlib. 22nd Jan 1659. "I will begin with Husbandery, about which you have been Theoricall & I have been Practical; but you will find that practice doth, and will give the law to Theoretics. I have tried divers of your experiments about bees and they signify nothing; like a chip in pottage, doth neither good nor hurt; but contending myself with the country-man rules & observations about them I find less trouble & more prifitt. M. Wall Causham" (Caversham Oxon). *Remind you of a beekeeper you know? Moses Wall communicated with Hartlib on many subjects*

Some relevant extracts from the University of Sheffield's digitisation of Samuel Hartlib's papers

and being a countryman often carried out experiments suggested by Hartlib, reporting back on the results. Not happy, but most of his other communications were of a similar tone – best leave it to us practical country people.

The Earliest Record of Beekeeping in Northern England

Appendix III

Firr deals

From *Rural Economy in Yorkshire 1641*, Henry Best, The Surtees Society, published 1857.

Firr deals were normally brought in from Norway; "reade-deale best". "Almost as durable as oak, and will not worme-eate soe soon as white deale ….. it is handsomer and better flowed".

They were 12feet X 12 inches, just over 1 inch thick, not shaken (cracked and flawed), not knotty, was the standard. They were priced from "£4 10s to £4 15s the hundredth". "There yoe six score deals to the hunderedth; and (for the most part) they putte just 1200 in one of theyre piles which are piled up in theyre yards". He gave very great details of the method used for 'piling'. Henry Best bought "200 reade deal of one Francis Taylor of Hull att £4 15s the hundredth" and gives details regarding the time taken to transport them from Hull and the number of deals per wagon. Common deal were 9d each. Both red and white deal was also available 14 feet x 14 inches at 12d – 18d and 5 groates each[41].

---

41 X built his wall to enclose his apiary "fower" yards high. Likely therefore that the construction was of 'firr deals' of 14 feet long.

The Earliest Record of Beekeeping in Northern England

Appendix IV

The octagonal bee-hive

From a member of his circle, Samuel Hartlib was made aware of the beekeeping exploits of the Rev. William Mew and his misleadingly labelled "transparent hives".

William Mew was educated at Cambridge University, admitted at Emmanuel College, Oct 1st 1618, Son of William of Eastington, Gloucester. B. A. 1622; M. A. 1626; B. D. 1633; ordained deacon July 22nd: priest 23rd, 1627. Rector of Easington 1635 – c. 1669.

William's son: -Samuel Mew, cler. fil. (Son of clergyman); Christ Church, Oxford University, matric. 14th May, 1651, demy. From Magdalen College 1653 – 4. B. A. 26th May 1654; Fellow 1654 – 69; M. A. from Magdalen Hall 18th Dec 1665 (as Meux.). Rector of Eastington 1669 – 1680, Canon of Wells 1680, of Winchester 1689.

William was chosen as a delegate to serve in the Westminster Assembly of Divines which was set up by the long parliament (1640 – 1648) to provide an authorised commentary on the scriptures to accompany the text of the English Bible which had been published in 1611. He was probably chosen because of his association with Sir Henry Vane, and as vicar of Eastington his patron was Sir Nathaniel Stephens, his M. P. and a Colonel for Oliver Cromwell. Stephens and Mew were both Puritans. The Assembly first met on 1st July 1643 and it was disbanded, having concluded its business in 1649. Mew was a constant attender and was chosen to preach the fast day sermon to Parliament on the 29th November 1643, which was later published; he was well known for his preaching.

Etiquette of the period required Hartlib to use an intermediary to make contact with Mew. The person he used was Nathanial Angelo, "Fellow of Eaton College" and member of his outer circle, who passed a letter from Hartlib to Mew in 1653 requesting details of his hive. Mew responded: -

"…. Since I left the hot service of the city, I have an apiary in the country………. The invention is a fancie that suits with the nature of that creature….. if you desire the model or description, I shall give the same to you that I did to Dr Wilkins[42]….who hath … set up one in his garden and ….. is setting up another with augmentations: I intended at first for an hyergliphick of labour."

Mew's letter also informed that he had studied Butler's book, and that the idea for his hives came from his personal observations and Pliny who wrote of a

---

[42] Dr John Wilkins was the master of Wadham, a position that he held from 1648 to 1659.

transparent bee-hive. He claimed that his 'transparent hives' had produced twice the amount of honey as others but never explained how or why, and it appears to have been based upon results from only two seasons. Mew stressed that the prime purpose of his garden erection was as example to his parishioners of the benefit of the work ethic and a well-ordered society. He added that a gentleman had bestowed a statue upon his 'phancie' to illustrate just that, but it succumbed in three years. In its place Mew had put 3 'triagonal' dials, 3 weather glasses, a clepsydra (time) and a cock (wind) – a weather station. He expressed the belief that his hive could increase the nations honey output by £300,000 per year, but only in the hands of someone as competent as him! Mew ended by admitting that previous to the letter, he had not been aware of Hartlib or his works, but invited him to visit or write.

The letter was dated 19th Sept. 1653 and written after Mew had returned from Westminster: it contains no evidence to support the commonly held belief that he gave Wilkins a transparent hive prior to his departure for Westminster, which was before Wilkins appointment at Oxford.

All very civil but it did not provide Hartlib with what he wanted, so on the 17th Nov. 1653, he responded with a very lengthy letter, from "my house near Charing Cross over against Angel Court". It commenced with much waffle then a request for "…a full description of your transparent hive, in the parts and dimensions thereof, and if you have any to spare (now the season is past), and would send up by the carrier…". He promised to reimburse costs and return the hive. Hartlib enclosed "a packet with several treaties and books" - his own. So now you know who I am, stop messing about! He wanted, or more realistically, get someone else, to test the hive in relation to the claims made by Mew for it.

Mew was not to be bullied, and responded with an even longer epistle: I have previously reported that he was considered an excellent preacher! He opened with a coded rebuke to Parliamentarians, a few of whom had formed the 'Anti-Pastoral Party' proposing that they should have pastorals of their own, Mew believed that this would leave him "stript to the bag and bottle", and indicated that he was now confined to his Hermitage, probably his way of expressing that he was living in religious seclusion with his parishioners rather than a description of his abode.

Mew followed by relating that when he left for Westminster he left a model of the hive ('an innocent phancie') in paste board[43] (closest comparison today is cardboard). His wife, who had died whilst he was away, had it made up in free stone (stone that can be carved) and set up in his garden. "In this I placed an

---

43   W. C. Cotton, in *My Bee Book*, 1842, noted that "Mr Drewitt – paper mill owner offers to make pasteboard hives at 6d each – he has a round one in use. So clearly not a new idea.

upper and lower hive"; taken literally this appears to say that he put a couple of boxes, arranged vertically inside the stone 'phancie'. Therefore, he did not do this until he eventually returned home in 1649. Hence, it is probable that the first time bees inhabited the stone erection was 1650. Mew was more inclined to describe the decorative features and their symbolism than give dimensions, and although he did give some information on his beekeeping management techniques, it was not very revealing.

Mew next attempted to answer Hartlib concerning the financial returns possible from honey. It was waffle, as one would expect, but he did include the information that he had heard that a gentleman of Norfolk made '300 li per year' – so effectively he had extrapolated from hearsay! He continued - "As for my transparent hives, I have but two, which are not moveable……" then "My apiary consists of a row of little houses, two stories high, two foot apart, which I find as cheap at seven years end as straw hackles, and far more handsome". He then states that he had unseen stalls in his bay windows in rooms that were used by lodgers. So not exactly the normal hermitage then!

His final paragraph expressed the hope that in future he would hope to communicate on weightier matters than Bees. He had tired of it, and was likely to ignore further requests about his bee-hive.

Sir your endeared friend, to serve you,

William Mewe (he appears to have used the 'e' here).

Eastington, Gloucester-shire, 20th December 1653.

But no 'Compliments of the season' and certainly not what Hartlib had hoped for! In effect it was a challenge to the authority that Hartlib believed he possessed by being in the employ of Cromwell.

Whilst the first two letters were complete copies, Mew's response was an extract, which begs the question – what must the rest of it have been like? Another sermon?

This exchange of letters conveyed little of value: -

- In 1650 Mew placed an upper and lower hive in his 'Phancie'. At no stage has he mentioned that they were octagonal, that is only derived from the Wren diagram. If they are 'transparent', the 'phancie' must also have been 'transparent' for the bees to be visible. This, of course, takes the writing of Mew literally, probably misguidedly so. He also states that he has only two 'transparent' hives, but he appears to have had only had one 'phancie'! Does this mean that the second augmented one was 'naked'?

- Mew indirectly informs that he set up his 'row of little houses' in 1646 - in the middle of his stint in London! This would have been possible because he did not spend 100% of his time at Westminster.[44]

- Mew repeatedly stated that the purpose of his 'transparent hives' was to provide an example to the members of his parish of the benefit of an unselfish work ethic, and thus reinforce the inscription on his 'phancie'. He had not intended the hives to be primarily honey producers which led to him having to use someone else's figures for possible national return. A misinterpretation by Hartlib?

Given his very confused writing and lack of definitive information, it is doubtful if Mew really contributed anything to the advancement of bee husbandry and eventually Hartlib lost patience with him, which was probably the preferred outcome for Mew. So Hartlib then turned his attention to John Wilkins whose knowledge of beekeeping was hardly better than that of Mew, but who was the possessor of a Mew style hive. Hartlib pestered Wilkins directly and through intermediaries who eventually asked his colleague Christopher Wren to send the information.[45] Hartlib's persistence eventually produced "A letter concerning that pleasant and profitable invention of a transparent beehive, written by that much accomplished and very ingeneous Fellow of All Soules College in Oxford, Mr Christ. Wren with the figure and description of said transparent bee-hive." There is no information on how this related to Mew's hive other than it was octagonal, and included a transparent piece of glass.

Wren wrote: -

Honoured Sir,

You have, by several hands, your desires to me of having a particular description of our three- storied bee-hive. I confess I was not over forward to execute this command of yours, and my reason was because the device, not fully answering our own expectation, I thought it would be much more unsatisfactory to you; but since you please to persist in your desires (as Mr Rawlinson told me the other day), I can no longer be shameless to persist in my incivility, especially prompted by mine own ambition, to find any way to show myself a servant to a person so eminent amongst the *ingenious* as yourself.

The description, I think is evident enough in the paper. I shall only tell you what effects we find. Last May, as I remember, we put in two swarms together, leaving the places to go in open only in the lower-most; but all the passage holes open from box to box: in the middlemost they first began their combs, then in the lowermost, before they had filled the middlemost, and so continued till they began to make two little combs in the upper box, (all this while deserted) and

---

[44] There is no record of when or if he returned home from Westminster during his time there, which questions the date of 1650 given for bees occupying the 'phancie'. I was unable to find Mrs Mew's date of death.

[45] The Mew/Wilkins contact was probably made through Mew's son, Samuel, who went up to Oxford University in 1651; I have found no other plausible link.

continued besides a part of a comb of the middle story, an inch or two up into the upper box, filling almost the passage hole quite up, leaving themselves only a little hole as big as two fingers might go in for their passage up and down. I am not very certain whether this was done at first when they wrought in the middle box; and whether this was not the reason why they wrought so little in the upper box, because they stopped themselves up from an easy passage to it. The combs in the lower story were well replenished with honey, and suddenly, but these little combs in the upper they quite deserted contrary to our expectation; which was that they would have wrought most in the upper story, and in the middlemost, in which, when they had wrought enough for their own spending, that then we might take away the uppermost from them, and so have continued still: but if we find another year that they fill not again the uppermost it will be all one still to take away the lower most from them; but if that be so, then two boxes will be sufficient. We must rather desire of you further light on this business, which I can presume you can afford us from the observations of others who have tried the like experiment for us, yet you see ours is imperfect.

Sir,

I am your obedient, humble servant,

Christopher Wren.

All Soules College

Feb. 16, 1654. (Some disagreement as to date – some give it as 26th).

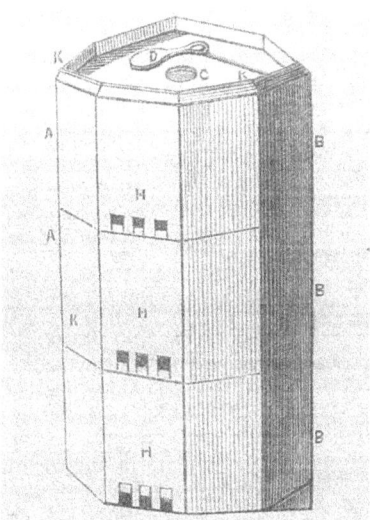

Mew/Wren hive

Wren and Wilkins rightly considered the first year with the hive to be a failure, but did not acknowledge that it was due to their lamentable lack of beekeeping ability.

Hartlib showed Wren's letter to a relative, Mr Greatrix[46], who had passed it to another (Sir Cheney Culpeper?). Both were of the opinion that because bees start at the top and work down, it was best to start with one box, adding a second underneath it when the bees had filled the first and same operation for the third box. "This I have gathered from such books as I have met with"[47]. When bottom box was fully worked out, the top box could be removed and another empty one added underneath. This individual was obsessed with making it easy for the bees – travel down when laden whenever possible, a pre-occupation of beekeepers at the time.

To summarise this: - Mew invents octagonal hive with glass panels in; passes information to Wilkins via his son, who has it made up; Wilkins and Wren were in the process of setting up another with augmentations. Hartlib obtains information from Wren. Because we do not have the original hive design of Mew, it is only possible to speculate what the augmentations were – possibly the lids with shutters.

Hartlib also included an entry on "some remarkable observations concerning the swarming of bees together with a short description of a bee-hive made of glass". No author is quoted but the similarity to the Mew hive in his bay windows is very noticeable. The hive description was very vague and I believe misleading, appearing to utilise large panes of glass. The section on swarming mentioned the part played by scouts (the word used), and the use of a "little den" to hold the queen in after the swarm was hived.

Some writers[48] have written eloquently about a meeting between John Streater, a soldier, pamphleteer and printer, and Mew. It is true that Streater served in Clonmel, Ireland, as a quartermaster and engineer for Cromwell, and that he would have passed close to Eastington on his way to Milford Haven. It is also likely that Streater printed Hartlib's Reformed Commonwealth of Bees, although his imprint does not appear on the title page; he certainly printed the *Reformed Virginian Silk Worm* which is often found bound with Hartlib's bee treatise. But that does not mean that they met; it is poetic licence. Streater had a very bumpy career.

John Evelyn (1620 – 1706), was born into a wealthy family enabling him to spend his time at leisure. He studied Natural History, and although this included Bees, Evelyn was not a beekeeper in the generally accepted sense of the word; he had a few hives of honey bees on his land but they were tended by retainers.

---

46 At Oxford a "College for Experiments Mechanics" was being erected where all the models of invention etc. were to be displayed. This was a similar to the Royal Society and its Repository. "They desire Greatricks to be the keeper of that College".

47 Unfortunately he did not share the information as to which books he had 'met with'.

48 Timothy Raylor and Adrian Johns.

He obtained his hives and knowledge of bees from others, and even for his time, much of his writings were incorrect, even the eminently rational W. C. Cotton later making fun of them. In 1652 Evelyn and his wife settled in Sayes Court, Deptford. It was here that he met Grinling Gibbons, a master woodworker/woodcarver on the estate, who he introduced to Christopher Wren and was probably responsible for the physical construction of the octagonal beehive that Wilkins had at Oxford. Evelyn was a founder member of the Royal Society in 1660, a diarist and a prolific writer on many topics.

John Evelyn was shown the octagonal transparent hive when he visited Wilkins at Oxford on 13th July 1654. His diary states that he was given "one of the hives" at that juncture, which he had erected in his garden at Sayes Court. The letter in Appendix I, dated 2nd April 1656, might be seen as contradicting this, because Wilkins mentions this to be the same as his at Wadham, and that he recommended Evelyn get two more the same, thus having the full three-story hive . Did Evelyn only get one box in 1654? If so, he needed two more, but if he had a full three-story job already, he would not need to be told by Wilkins. It is known that Evelyn wrote his diary up retrospectively from notes made at the time, and the accuracy of some entries has been questioned. Evelyn painted the inside of his hive with rosin, acknowledging that some others used matting. The purpose was to aid the bees in attaching their hives to the box, not for insulation or protection. However he still 'dressed' his new hives.

Evelyn set himself the task of writing *Elysium Britannicum,* an account of all things natural, but it is doubtful if the task was ever completed, certainly only 342 pages of manuscript survive, of which it is believed was intended to be 900 pages in total. By advertising the index for the complete work, he hoped that others would complete some of the sections that were in their area of expertise. In a letter to Sir Thomas Browne of Norfolk, Evelyn wrote of the MS: -

…… to refine upon some particulars especially concerning the ornaments of Gardens which I shall endeavour so to handle that persons of all conditions and faculties which delight in Gardens, may therein encounter something for their own advantage.

This was a work was intended for those with deep pockets and large tracts of land. It was an assemblage of the horticultural knowledge of the time, and as such much was taken from others: with the rapid development that was taking place in this period, it was always likely that he would never finish it, always refreshing parts as new revelations came to light. Even so one recent reviewer of the work commented that: -

His solemn discussion of the spontaneous generation of insects similarly suggests that he was not at the cutting-edge of contemporary entomology.

Fortunately the bee material is in the surviving papers, and contains a pen and ink sketch of "a transparent Bee-Hive", but it is not that of his initial acquisition by Mew/Wilkins/Wren; it was intended to be Rusden's hive. This is covered more fully by D. A. Smith, in his excellent Bee Research Association booklet, *John Evelyn's Manuscript on Bees from Elysium Britannicum*[49], published in 1966. Evelyn refers to: -

Samuel Hartlib, *The reformed Commonwealth of Bees*, 1655;

Plot, *The Natural History of Oxfordshire*, 1877;

Rusden, *A Further Discovery of Bees*, 1779;

In the matter of bees and beekeeping Evelyn was similar to Hartlib, in that the bee part of his manuscript was his write-up of other author's works, and in some instances it is a direct copy of material sent to him. However, unlike Hartlib he did have bees on his estate.

John Evelyn in his famous diaries, described Hartlib as "Master of innumerable curiosities and very communicative". Even though they supported opposite sides in the Civil war, in the late 1650s, Evelyn had shown Hartlib the part of his *Elysium Britannicum* manuscript that he had finished at that time, and they were regular correspondents.

Evelyn started by detailing the ancient writers that extolled the virtues of bees, but he still referred to the big bee as King, and made much of bee society as would be expected. He followed with a contributed passage which owes much to Southerne, but is not credited.

As he meanders through his manuscript, Evelyn often has contradictory material which indicates a lack of editing. The shape of the hive is one such instance. He states that hives can be any shape, but hexagonal "seems to be the most agreeable because it resembles the shape of their cells". Earlier he expressed the opinion that a cell was that shape because it had one angle for each foot of the bee! However, his illustration is of an octagonal hive as Mew. The volume of each box he gives as 3 pecks (= 0.96 cu. ft.), and is best made from Oak wainscot or clapboard (panels). As stated earlier he recommended coating the inside with Rosin, a resin extracted from pines, rather than using rush matting as Wren did which implies that neither had an internal frame. The purpose was to improve the attach-ability of the combs to the boxes. The outsides were to be painted with white oil and decorated as desired. All joints were dovetailed.

---

49   Also see complete transcription *Elysium Britannicum or the Royal Gardens*, John F. Ingram, Pennsylvania Press.

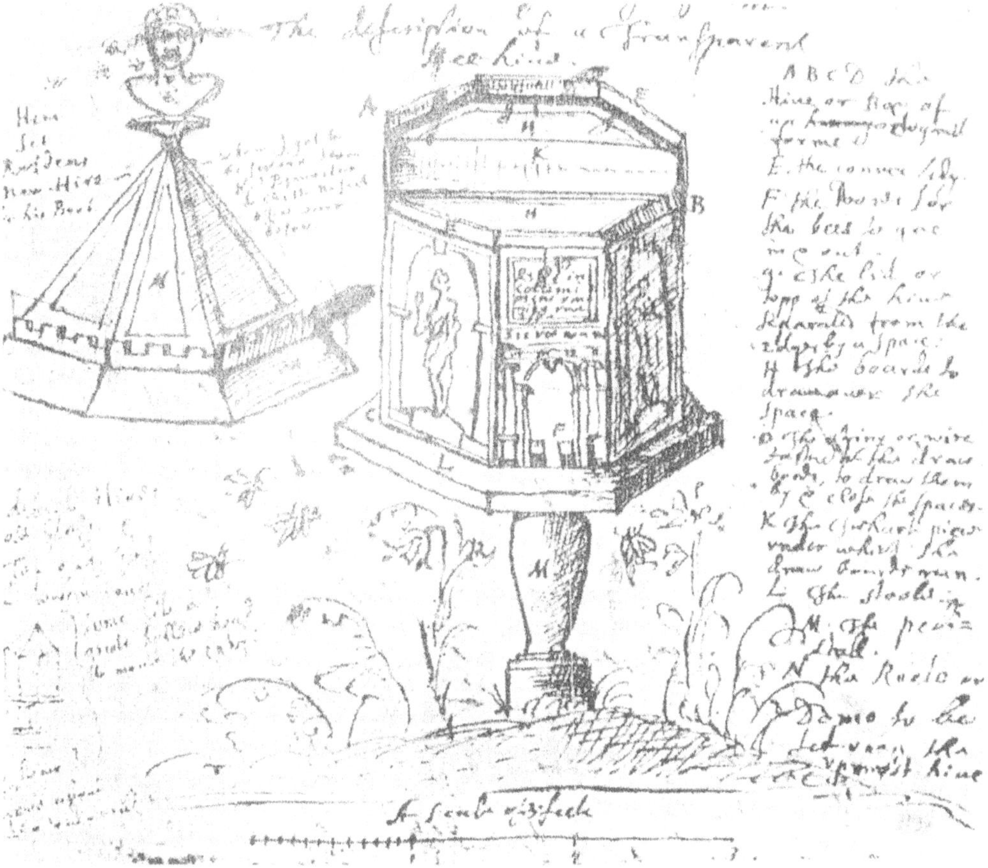

The octagonal bee-hive

Evelyn's hive

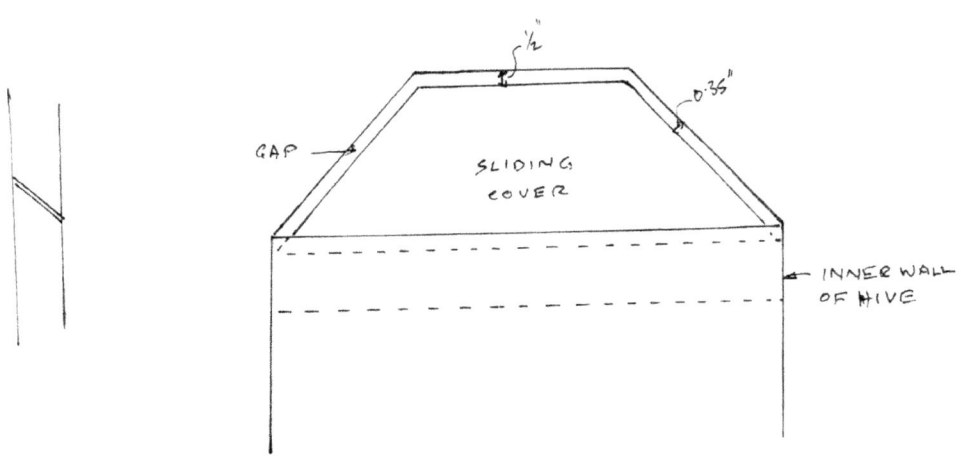

Evelyn hive, joints and closing cover

In his description, having detailed the box Evelyn wrote: -

Of these (boxes) you shall provide store, made exactly of the same shape, with a flat bottom or lid, which yet shall be fitted as that two parts thereof be separate from the sides of the Box almost half an inch, for a reason to be hereafter showed; and these spaces shall also have covers running in two grooves, to be drawn over them by wires at pleasure. …..the upper part of the box must advance above the lid a full inch, to be sloped away convexedly, the bottoms concaveley.

Each box was equipped with at least one 6 inches x 4 inches window and hinged cover. 1 inch below the top of each box was a lid which appears to comprise one permanently attached board and two sliding pieces which move below the fixed board. He does not give constructional details to show how this was to be achieved but it would have required a slot to be cut into the two opposite sides for the slider to run in, or an additional support under the slider on those two sides. Neither of these suggestions would provide support of the sliding panel at its narrower end.

The slider was to open half an inch. If this is assumed to be the distance from side B to the slider, the perpendicular distance for the other two sides would be 0.35 inches. The question occurs – why did Evelyn choose these dimensions? He promised an answer later, but it never appeared. Of course, both dimensions are close to 'bee-space'. It is clear from his description that the sliders were only to isolate the boxes one from another and not to cut through comb so he did not anticipate comb being built through the gap or used wire as earlier. Propolis would however have been a consideration in this sliding mechanism and in the joints between the boxes. The roof was fitted with the concave vertical parts to fit onto the top box, and overhung the box by 3 inches; the stand was 2 inches wider all round than the hive, so the water that ran off the roof should have fallen onto the ground and not the stand. The stand was two feet high.

The image of Evelyn's hive included a scale and using this, side to side of the octagon is 18 inches and height 15 inches. Assuming timber 1 inch thick this results in a volume of 2860 cu. inches. One bushel is approximately 2220 cu. inches. and with one bushel equivalent to four pecks, the hive illustrated has a volume of about 5 pecks; not the volume he was trying to achieve. Whilst the illustration is clearly a free hand sketch it brings into to question the credibility of the other information being presented. One might be led into thinking that the illustration is of the outside box of a double-walled hive; but this is unlikely because the isolating mechanism is illustrated covering the whole of the box shown.

Evelyn started by putting a swarm into one box. When the bees had worked this out as witnessed through the window, another box was inserted under it and the entrance in the original box was closed. When both were worked out, the boards were closed and the top box removed. An empty box with its top boards open was then placed under the remaining box which had its entrance closed. The removed box is emptied of bees[50] and the honey harvested. He acknowledged that others operated variations on this – three boxes before removal or adding boxes above rather than below. I believe this to be less good than X. He correctly credited Mewe with the design adding "…. The happy product of his exile or eclipse during our unnatural wars" which is incorrect. Evelyn then dealt with Rusden's hive, in a section which must have been written after 1679, and confirms that he was responsible for Rusden being appointed the Kings bee-master, much derided by others. Rusden's hive was octagonal, 16 inches wide, 10 inches high which results in a volume close to that which Evelyn said he was aiming for.

The Evelyn manuscript continues with more from Hartlib – Brown's hive; the Dutch recumbent hive, both aiming for and achieving the volume of a bushel. Evelyn expanded Hartlib's Italian entry by extracting information from Vincenzo Tanara, *Oeconomia del Cittadino in Villa (The economy of the city in villa)*, 1644. Unfortunately it is somewhat confused. He states that the hives are two feet long and 8 inches in diameter, indicating recumbent hives, but when describing driving to 'gelt' the hives he writes "…. drive them by setting an empty hive upon the full ….." which indicates that the hive was vertical, but given the measurements not very stable one would guess. He drove by holding the combination over smoke, drawing a board between the individual cylinders to prevent the bees escaping whilst the honey is taken. The dimensions produce a hive of just over half a bushel, which leads me to question the figures, but he also reported that Tanara claimed that bees placed in cellars over winter would survive on the scent of the wine!

There is nothing original in Evelyn's work, and much reflects X's manuscript, including material about migratory beekeeping; on choice of bees when buying "… choose those who are of a bright brown, smooth, and of middling size", and to treat "punctures and stinging of bees cured by their own honey, by juice of malows, by cow dung mixed with vinegar". This is not intended to imply that Evelyn and X had met or communicated in any way, social etiquette of the period made that highly unlikely.

Evelyn's account of beekeeping in the 17th century is a regurgitation of information gathered from others and most of it already in the public domain, indicative of someone who wrote about topics with little personal involvement. He did no gardening or beekeeping, simply employed others who did. There is nothing in this manuscript which detracts from the importance or originality of the work by X, or adds to our knowledge of beekeeping at this time. X made no

---

50   No information on how was supplied.

reference to Evelyn or his manuscript, although he could have included them in the second part.

Observations
- I cannot find Evelyn's later reference as promised, regarding the reasoning behind the distance between the sliding 'cover' and the side of the hive, which is a shame because the right angled distance between shutter and side of hive varies between "almost half inch" and 0.35". Tantalising hovering either side of BEE SPACE.
- There is the potential problem with propolis in the joint between boxes and the sliding mechanism.
- The sliding shutters restrict bee movement box to box. Some beekeepers use a similar method today.
- Rather intricate carpentry was required given the period, any warping of the closing board would prove awkward and keeping it free to move easily, difficult.
- The Evelyn pen and ink drawing in Elysium Britannicum, is meant to be based upon Rusden's hive, but the arrangement of the slider is different. It is not possible to date it.

Henry Oldenburgh, who with Wilkins was to become joint secretary of the Royal Society at its formation, also saw the hives at Oxford and wrote: -

…..here in Dr Wilkins garden two very fine glass hives which I am very much taken with both for ye entertainment this invention giveth to ye mind about the curious workmanship of this creature, and then the mercy which thereby the same may receive for its busy pains.

Oldenburgh was very much part of the Hartlib/Dury axis; he was a friend also of Robert Boyle and tutored his nephew, Richard Jones. His second wife was the daughter of John Dury. He is generally considered to be the initiator of peer review. Upon Hartlib's demise he took over his network of contacts.

From the entries in his treatise after his release from gaol, it appears that X had also met with Wilkins. However, I do not believe that he saw the transparent hives, otherwise he would have written of it. This implies that he visited Wilkins between the time of his release and early 1653 – probably immediately after re-gaining his freedom.

Another visitor to Oxford was Robert Wood, who wrote to Hartlib:

Dr. Wilkins Bee-hive is accurately described in your treatise of bees, and for its transparency tis caused by a piece of glass a little bigger than my hand set into the hive on one side thereof.

Illustrating that despite the hype, the window was of little value. In Evelyn's hive the window was 6" x 4", little better. Larger panes of glass were not available until 1670.

Traditionally historians have always linked the hives of Mewe, Wren, Gedde, Rusden, Evelyn, Thorley and Stewerton, because they were all octagonal boxes, with or without some internal structure.

John Gedde (y,ie)

John Gedde, born about 1630 in St Andrews, Scotland, where his father was Baillie. After an apprenticeship with William Henderson, the writer[51] to the signet, he became writer in Cupar. When Charles II was in Scotland trying to raise money and an Army, John although still apprenticed, became clerk to the committee of war, for the period 1650/1. He made a large sum of money in those two years, after which he was involved in many land and other deals. His professional career was severely hampered because of his religious convictions – he was a Covenanter, and refused to comply with 'the English' and 'subscribe to their oath called the tender'. His dealing resulted in getting involved in many disputes which occupied much of his time and appears to have lost him much money. In 1665 he was appointed Baillie of Falkland and the following year he travelled for the first time to London.

Gedde claimed that he "fell upon new experiment and discovery of better ordering and improving bees" in 1668 and that the stimulus was the amount of honey bees collected in 'his Hives'- in trees. The only evidence of this is from Gedde's own hand; on the title page of the first edition of his book *A New Discovery of an excellent method of bee-houses and colonies,* published in 1675, where he also states that "Experienced seven years by John Gedde, Gent. Inventor ….".

Gedde supplied details of the hive, and possibly a sample, to Sir William Thomson and others; Thomson passed it to Sir Robert Murray (Moray), and in 1672 Murray presented one of the Gedde hives to the Repository of the Royal Society[52]. He called it "a bee-hive of a peculiar contrivance" adding it was "made up of several pieces to take off one, whereby bees are kept from swarming by adding a new box for every swarm". Sir Robert was not a beekeeper, and the message could well have become scrambled as it was passed along the line. A year later joint secretary of the Royal Society and a senior member of the Hartlib circle, Henry Oldenburg, wrote an account with illustration of Gedde's hive, in the Philosophical Transactions of the Royal Society. Neither Murray nor Oldenburg mentioned Gedde in their written pieces, probably believing that Thomson was

---

51   A Scottish solicitor at this time.
52   The Repository was nothing more than a room at Gresham College, home to the Royal Society at the time, where inventions were displayed. Reports indicate that it was a shambles.

the originator. The hive was described thus: -

"The bee-house is made of wainscoat, about 16 inches in height 23 inches in breadth between opposite sides. It hath 8 sides each almost 9 inches in breadth. It is close covered at top with boards having a square hole in the middle, 5 inches long and about 4 inches broad, with a shutter that slides to and fro, in a grove ½ inch longer than the hole. It has two windows opposite one another, and may have more of any figure, with panes of glass and shutters. The doors for the bees is divided into three or four holes about half an inch wide and as high, with a shutter that slides in a grove to cover them in winter. It hath two iron handles, with joints to be placed about the middle if there be no windows on the sides where they are; or above them if they be. At the top it hath a crease all all round it, about ½" depth on one side, and 1 ½" high: and another on the inside at the bottom, which serves to fix them when upon one another. It hath also a hole about two inches in height and as much in breadth, on one side at bottom, by which the knife is put in to cut the bees work, that passes through the whole, from one bee-house into another, as they work downwards into the empty house, which hath a sliding shutter to cover it. Within the bee house there is a square frame made of four posts joyned at the top and bottom, and in the middle , with four sticks for the bees, to fasten the work upon, which, though they will serve, yet it may be securer to have two more added in every one of these places, crossing the frame either from the middle of the opposite side sticks, or from angles, where the posts are placed."

John Gedde became aware of the Royal Society's publication within a year of its appearance. Not happy with the leaking of details of his bee-hive without acknowledgement, he could envisage that any financial gain from it would not be his, so he travelled to London, took partners/agents and obtained a patent for it from Charles II. It would not have proved difficult, Charles was at least aware of Gedde following his time in Falkland, and at this time, there was no proof of uniqueness required. His initial partners were William Galt, Gent. London, and Samuel Nowell, Gent. (probably the solicitor) who was soon replaced by Thomas Douglas of London, doctor of physic. By 1677 William had been replaced by John Galt. The agents were necessary because of Gedde's refusal to swear the English oath. They were soon joined by Thomas Blond of Westminster, who appears to have been given the duty of overseeing Gedde's interests in England and he was later he was replaced by Moses Rusden. The patent was awarded "for their invention of such commodious hives and houses……..."; showing that they were all allegedly responsible for the invention and that Gedde had the bee houses included. The rapid changing of partners might well indicate that it was not proving very lucrative.

Gedde's partners advertised in the London Gazette, for agents to cover all areas: they were to manufacture the hives and sell or use them under licence. The margins were only 12½%, and each licence holder was required to purchase a sample hive for £1. Andrew Paschall, rector of Chedzoy, Westonzoyland, Somerset, approached John Howe, a barber/surgeon[53] of Bridgewater, suggesting he might become the Gedde agent for Somerset. Howe was not convinced that the invention would 'take' in the country, so Paschall[54] asked antiquary and gossip[55], John Aubrey for his opinion. Paschall wrote: -

I observe that the chief market for bees is among the middle and meaner sort of people who are not likely to be willing at so great a charge… The Gentleman may make a cheaper contrivance nearer the common way of keeping them in straw hives….. and where (sic) such a thing described In print and cheap licences granted (I suppose 2/6 price)[56] he might serve the publicke and his own profit more and yet continue to offer persons of station at his own rates.

This would appear to indicate Aubrey was postulating a square/rectangular tiered straw hive with an inner frame. This was long before such a hive appeared.

In 1675, John Gedde published the first edition of his book *A New Discovery of an Excellent Method of bee-Houses and Colonies.* As the title indicates Gedde had added a bee-house. The book was a very small volume of 30 pages and 1 plate, and intended to complement the purchase of a hive, providing details of the system of management for it. It was followed the next year with an enlarged volume in which Gedde attempted to answer questions resulting from the first edition. A third edition followed a year later again. The second edition was accompanied by a single sheet 'flyer' describing how to correctly use the hive, which would appear to negate the purpose of the book. In all editions of his book Gedde claims that it was "approved by the Royal Society" but nothing could be further from the truth, as with the statement that it was "the best form of hive yet invented", which he claimed the Royal Society had reported. The later editions went further in the body of the books stating "The approbation of the Royal Society and of the most famous Bee-masters in England", adding that the "experience of many persons of quality and many others of rank, with the approbation of his Royal Majesty, is a sufficient testimony of the commodiousness and benefit of this new invention".

Nothing that the members of the Royal Society wrote of the octagonal hive

---

53 This would have made him a Barber/surgeon/beekeeper/ hive manufacturer and agent!
54 Andrew Pascall was a fellow of Queens College, Oxford from 1653 to 1662 when he became Rector of Chedzoy until his death 1696. He is best known for his account of the Battle of Sedgemoor, which took place adjacent to his parish. He was a regular correspondent of John Aubrey.
55 Not a derogatory term in the seventeenth century.
56 The account books of Sir John Foulis a contemporary of Gedde, who lived close to Edinburgh, listed skeps with bees being purchased in the late 1600s for about eight shillings each.

mentioned Gedde or that anything about his hive was novel; we know that at best the only originality was the frame, and surely if the Royal Society knew this they would have said it; their purpose was simply to publicise it.

The Royal Society are said to have tested the claim that the hive stopped swarming and found it to answer the design but there is no written evidence that any member of the Society had carried out any trials, and what does 'answer the design' mean? However, Charles II was sufficiently impressed to have one erected in Spring Garden Whitehall and one at Windsor, where he watched the bees. The King also commanded that one of the hives be erected in his park at Falkland in order that it would stir noblemen et al, to install them also. He gave John Geddy 20 aikers at the end of his park in Falkland to be enclosed, trenched and planted suitable for bees. Gedde was also granted £200 sterling to erect a house on this land and accumplish the set task. There is no evidence to prove that Gedde achieved any of this. As there is so little difference between Gedde and Rusden's hive one illustration of Rusdens will suffice.

In order to quell the howls of objection to his patenting a hive of which only part could in any way be claimed as original, Gedde in the second edition of his book, included a letter from Samuel Mew that he had received: -

Worthy Sir,

I have with great delight perused the directions you have given to the public concerning your new invention ……. I have had for these twenty years and upwards the opportunity and content to observe these pretty artists you contrive for, at work through glass placed for that purpose …..

Samuel then detailed the difficulty that he had in determining when to remove the top box to take the honey. He continues: -

I congratulate you [on] your happy invention, and the perfection it is already reached to, the Royal society's approbation and you own seven years experience, I am the more reall herein in regard I have formally imployed and wearied my own thoughts on this subject and was fain at last to give it over as unfeasible; partly from the inconvenience above said and partly for the want of an apprehensive (?) workman: but now I am where I would be at anothers trouble, and may enjoy the fruits of your pains at an easy rate. I heartily wish your profit may pay your labours both as you are an inventor and a Beemaster. If sir, your occasions will suffer, and you think fit to favour me with a line or two, be pleased, I pray, to send the price of one of your boxes with the frame, as also a licence from yourself to make use of it, and you will yet further engage your unknown but real friendand servant,

Samuel Mew.

Gedde responded: -

...and for your further satisfaction I have sent with the Gloucester carrier, one colony of boxes, one licence, and a book of direction s, to be used by you, as a token of my respects to a person of so much worth; hoping that after you have received and viewed the Colony, there will rest no place for any futher scruple.

Even given the language and protocal of the time, a very strange exchange. Gedde excising his conscience? Why would Samuel have such a question if, as he says, Gedde had already published the directions for use of the hive? It does however, reinforce the date that William Mew had the hive with windows, (1675 minus about twenty years equals approximately 1655), and that William was now dead; believed to have died in 1669.

Moses Rusden published the first edition of his book *A further Discovery of Bees* in 1679; his hive was similar in most respects to that of Gedde, and he acknowledges that it was a 'development' of that hive. Rusden was not shy either, the title page of his book claimed: -

By Moses Rusden, an Apothecary; Bee-Mafter to the King's most excellent Majesty. Published by His Majesties afpecial Command, and approved by the *Royal Society* at *Gresham* Coll.

Rusden's hive was an octagonal box, 10 feet high, 16 inches across, outside; assuming 1 inch thick timber gives an internal volume of 1620 cu. inches. One imperial bushel = 2220 cu inches. Each box had a fixed board across the top with a 5 inches square hole and a slider to close it when required. There were windows front and back, and an internal frame to support the comb. He preferred Firr "because it sucks up the breath of the bees in cold weather".

To transfer a colony from a skep, it was placed over a single box, top hole and entrance open. The bees had to pass through the empty box to get in and out, and would work it out[57]. When fully worked out as observed through the glass, he added a second box underneath, again leaving only the bottom entrance open. When that one was fully worked out, the shutter in the top box was closed and the skep removed. A third box was added under the other two and when that had also been worked out the top box can be removed and the honey taken. There is nothing novel here by way of a hive but he also described a bee house, just a box 10 feet long, 2 ½ feet deep and 3 ½ feet. high with doors to the front and back. It was intended to hold four hives. Again it was not novel, Remnant amongst several, had been there before.

---

57   This technique is still in use by beekeepers who 'take' swarms in skeps.

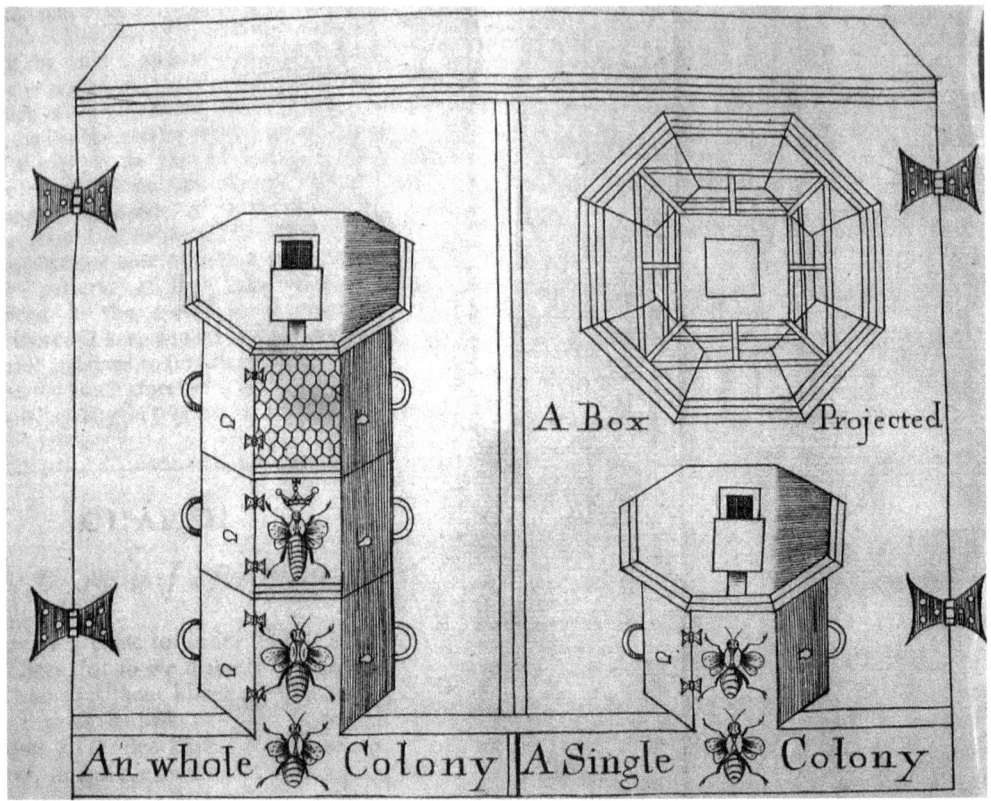

Rusden/Gedde hive

D. A. Smith resolved without doubt that Rusden was entitled to the title 'Kings bee-master' but the lack of information to date the certification which appeared on Evelyn's sketch of his hive, does not inform as to when the title was conferred but it must have been before 1679 when Rusden's book first appeared.

Gedde's life continued in the same somewhat disjointed and controversial manner. Continual problems arose because of his religious beliefs which appear to have resulted in him and his wife living in the North of England for an extended period, and he was charged for attending an illegal meeting in London. Additionally there was considerable upset when John broke into a new burial ground where his daughter, who died age 16, was buried, and put a rail around her grave.

Could Gedde have met X whilst in Northumberland? Lack of X's date of death hampers this line of thought, but it is unlikely.

I could find no genealogy information on John Gedde, but there is information that he was still alive in 1695, when he was back in Falkland, and listed as a heritor – property owner liable to local tax. The picture is clouded when in 1697 the second - third edition of his book was published and although it still gives Gedde as the author, it appears that Richard Swalden, significantly a joiner, was now in charge of Gedde's affairs in England, taking over from Rusden presumably, although there are no definite dates for any of this.

So what can possibly be made of this?

- Both Gedde and Rusden improperly claimed that their hives had the approbation of the Royal Society, this assumes that Rusden was claiming his hive was different than Geddes. Can either be trusted?
- Rusden's book was a much better effort than that of Gedde, but neither had the detail or was as advanced as X's manuscript.
- The Royal Society did not mention the unique nature of the hive, indicating, perhaps, that there was none.
- Neither the management detailed by Rusden or Gedde was as advanced as that in X's manuscript, only part of which X claimed to be original, several years later

It is not uncommon to have two beekeepers in conflict over ownership of beekeeping inventions, in the late 19th century the Renfrewshire beekeeper and the Ayrshire beekeeper laid claim to most things new, they also wrote much on the Stewarton hive, an alleged development of the Mew/Gedde/Rusden/Evelyn hives. However, before getting over excited and raising the blood pressure further, one should remember that it is just an octagonal box that beekeepers kept honey-bees in.

Hive capacities

Gedde wrote: - "For their dimension, it is fit that each box be of that largeness, that it may contain a bushel, the height not exceeding 12 inches, and its breadth one third more than the heighth."

Rusden gave height and breadth measurements.

The Royal Society gave dimensions of height, overall breadth and width of each panel.

Evelyn wrote that each box should contain 3 pecks, but with no measurements given we are left to derive them from his sketch and scale, never likely to be accurate. Also for an individual who stated in his text regarding shape, that "… the hexangular seems to be the most agreeable because it resembles the form of their cells", and then proceeds to show a sketch of an octagonal hive, it is not likely to be reliable either. Many writers on bees at this time believed the bee cells were hexagonal because the bee had six feet but of course never managed to logically link the two.

There are no records giving the size of the Mewe hive.

Wren did not give dimensions and again we have to resort to measurements from a drawing.

My calculations give the following approximate volumes for the various hives: -
Wren (Mewe), from Hartlib – 4600 cubic inches.
Evelyn – 4200 cu. ins.
Royal Society (Gedde) – 4900 cu. ins.
Gedde – 3000 cu. ins.
Rusden – 2500 cu ins.
[1 bushel = 4 pecks = 2150 cu. ins.]
Notes: -

The Royal Society measurements were presumably from the sample hive in the Repository, it is difficult to see how these could be wildly out. However, did Gedde supply the hive or did Thomson have it made up, if so from what information and how accurately?

Wren and Evelyn's capacity both arise from scaled sketches; Wren's having the additional interface of the engraver/printer. Even so it is difficult to imagine how Evelyn could have been so far removed from the aimed for figure. If Evelyn's hive is double walled as some believe then the sketch showing the closing mechanism on top of the box is wrong/inaccurate because it is shown closing onto what would have been the outer wall.

There appears to be a distinct division between the results before the Gedde patent and after it.

Appendix V

Miscellaneous

Hartlib was friendly with Walter Blith a noted writer on husbandry. Sir Kenelm Digby was also a friend. Hartlib used Sir Cheney Culpeper as a conduit to parliament for agriculture matters.

Hartlib's son (Sam) was employed as London agent for Berwick on Tweed and was appointed as a solicitor for the Merchant Adventurers of Newcastle upon Tyne.

The Earliest Record of Beekeeping in Northern England

References

Manuscript, *A Treatise of bees*, James Marshall and Marie – Louise Osborn collection, Beinecke Rare Book3 and Manuscript Library, Yale University.

Collectif, *Histoire de l'agriculture gauloise gallo-romaine et medieval,* 1596 reprinted 2006.

My own notes from the *Memoir of John Geddy* in the Miscellany of the Abbotsford Club. Hathi Trust Digital Library.

Article on John Geddy by H. J. O. Walker based upon Miscellany of the Abbotford Club. *British Bee Journal*, 1922; pp 265, 306, 401, 438, 521.

D. J. Bryden, *John Gedde's Bee-house and the Royal Society,* Notes Rec. Royal Soc. London, 1994.

D. J. Bryden, *entry on John Gedde in Oxford Dictionary of National Biography.*

John Gedde, *A new discovery of an excellent method of bee houses and colonies.* 1675.

Moses Rusden, *A further discovery of bees. 1679.*

Joseph Warder, *The True Amazons. 1712.*

John Worlidge, *Apirium. 1676.*

Samuel Hartlib, *The reformed commonwealth of bees.* 1655.

Samuel Hartlib, *His legacy of husbandry……* 1655, EEBO edition 2011.

The Hartlib Papers Project – University of Sheffield.

M Greengrass, *entry in Oxford Dictionary of National Biography.*

E Crane, *World history of beekeeping,* 1999.

D Woodward editor, *The farming and memorandum books of Henry Best of Emswell, 1642 – 1684.* The Surtees Society, 1857.

Timothy Raylor, *Samuel Hartlib and the Commonwealth of bees,* in Michael Leslie and Timothy Raylor, *Culture and cultivation in early modern England,* 1992.

Adrian Johns, *The nature of the book,* chapter *John Streater and the Knights of the Galaxy,* 2000.

M J Braddick & Mark Greengrass, *The letters of Sir Cheney Culpeper 1641 – 1658.* Camden Miscellany XXXIII, 1996.

Vincenzo Tanara, *Oeconomia del Cittadino in Villa (The economy of the city in villa)*, 1644.

Louis Mendez de Torres, *Tractado breve de la altevation y uera de las colmenas*, 1586.

Rev. Charles Butler, *The feminine monarchy*, third edition 1623.

John Levett, *The ordering of bees*, 1634.

Richard Remnant, *A discourse or historie of bees*, 1637.

John Evelyn, *Elysium Britannicum*, 1655. Manuscript on www.

Sir William Thomson, *A description of a bee-house*, Trans. royal Society, 1673.

Robert Plot, *A natural history of Oxfordshire*, 1676.

John Britton, *Memoires of natural remarques in the County of Wilts by John Aubrey*, 1847.

www.ingramcontent.com/pod-product-compliance
Lightning Source LLC
LaVergne TN
LVHW081355060426
835510LV00013B/1834